Studies in Systems, Decision and Control

Volume 313

Series Editor

Janusz Kacprzyk, Systems Research Institute, Polish Academy of Sciences, Warsaw, Poland

The series "Studies in Systems, Decision and Control" (SSDC) covers both new developments and advances, as well as the state of the art, in the various areas of broadly perceived systems, decision making and control–quickly, up to date and with a high quality. The intent is to cover the theory, applications, and perspectives on the state of the art and future developments relevant to systems, decision making, control, complex processes and related areas, as embedded in the fields of engineering, computer science, physics, economics, social and life sciences, as well as the paradigms and methodologies behind them. The series contains monographs, textbooks, lecture notes and edited volumes in systems, decision making and control spanning the areas of Cyber-Physical Systems, Autonomous Systems, Sensor Networks, Control Systems, Energy Systems, Automotive Systems, Biological Systems, Vehicular Networking and Connected Vehicles, Aerospace Systems, Automation, Manufacturing, Smart Grids, Nonlinear Systems, Power Systems, Robotics, Social Systems, Economic Systems and other. Of particular value to both the contributors and the readership are the short publication timeframe and the world-wide distribution and exposure which enable both a wide and rapid dissemination of research output.

Indexed by SCOPUS, DBLP, WTI Frankfurt eG, zbMATH, SCImago.

All books published in the series are submitted for consideration in Web of Science.

More information about this series at http://www.springer.com/series/13304

Józef Korbicz · Krzysztof Patan ·
Marcel Luzar
Editors

Advances in Diagnostics of Processes and Systems

Selected Papers from the 14th International
Conference on Diagnostics of Processes
and Systems (DPS), September 21–23, 2020,
Zielona Góra (Poland)

 Springer

Editors
Józef Korbicz
Institute of Control
and Computation Engineering
University of Zielona Góra
Zielona Góra, Poland

Krzysztof Patan
Institute of Control
and Computation Engineering
University of Zielona Góra
Zielona Góra, Poland

Marcel Luzar
Institute of Control
and Computation Engineering
University of Zielona Góra
Zielona Góra, Poland

ISSN 2198-4182   ISSN 2198-4190   (electronic)
Studies in Systems, Decision and Control
ISBN 978-3-030-58966-0   ISBN 978-3-030-58964-6   (eBook)
https://doi.org/10.1007/978-3-030-58964-6

This Springer imprint is published by the registered company Springer Nature Switzerland AG
The registered company address is: Gewerbestrasse 11, 6330 Cham, Switzerland

Preface

The continuous increase in the complexity of modern industrial systems, as well as growing reliability demands regarding their operation and control quality, are serious challenges for further development of the theory and practice of control and technical diagnostics. Nowadays, technical diagnostics is a field of intensive scientific research and covers well-established topics, as well as emerging developments in control engineering, artificial intelligence, applied mathematics, and statistics. At the same time, a growing number of applications of different fault diagnosis methods, especially in the electrical, mechanical, chemical, and medical areas is being observed. The rapidly increasing complexity of automation in the industry and the continuing need to ensure reliability and safety at the highest level require ongoing research and development of innovative fault diagnosis approaches. Furthermore, special attention is paid to fault-tolerant and self-reconfiguring control systems, which are crucial wherever maintenance or repair cannot be realized immediately.

The subject matter of the conference was replying to the expectations and demands of research and industrial centers related to modern and safe diagnostics systems, process monitoring systems, and expert systems. In general, conference topics correspond to the subject area of the IFAC symposium on *Fault Detection, Supervision, and Safety for Technical Processes (SAFEPROCESS)* and are extended to some issues of medical diagnostics.

The book contains selected papers presented at the 14th International Conference on *Diagnostics of Processes and Systems, DPS 2020*, held online in Zielona Góra, Poland, on 21–23 September 2020. The conference was organized by the Institute of Control and Computation Engineering of the University of Zielona Góra in cooperation with the Warsaw and Gdańsk Universities of Technology. The history of the *DPS* conferences dates back to 1996, and has been organized every two years sequentially by the three universities (in Zielona Góra, Warsaw, and Gdańsk). The *DPS* conferences attract each time approximately 100 participants, as well as internationally recognized plenary speakers, including so far Paul Frank,

Ronald J. Patton, Tadeusz Kaczorek, Jakob Stoustrup, Jan Maciejowski, Eduardo F. Camacho, Christopher Edwards, Steven Ding, Didier Theilliol, Ryszard Tadeusiewicz, Vincenc Puig, Paolo Castaldi, and Jérôme Cieslak, among others.

The aim of this collective book is to show the bridge between technical and medical diagnosis based on artificial intelligence methods and techniques. The volume is divided into three parts:

  I. Fault-Tolerant Control and Reconfiguration
 II. Fault Diagnosis of Processes and Systems
III. Medical Applications

and consists of 13 carefully selected full papers. Every submission was subjected to a thorough peer review process (2 to 3 independent reviews per paper), and only those with a consistent and strong recommendation from the reviewers have been accepted and included in the proceedings.

In this regard, we would like to express our gratitude to the authors for their valuable submissions and to the members of the International Program Committee, who served as the reviewers, for their inestimable effort put into the evaluation process. We believe that this book will become a great reference tool for scientists working in the area of technical and medical diagnostics.

Zielona Góra, Poland   Józef Korbicz
June 2020   Krzysztof Patan
                                                                                      Marcel Luzar

Acknowledgments We would like to acknowledge the technical and administrative support of Kinga Włoch (financial manager), Ewa Lehmann (organizational manager), as well as Paweł Kasza and Andrzej Czajkowski (web management and on-line support) for their commitment and effort applied to make the conference a very successful event.

Contents

Medical Applications

Fault-Tolerant Control
and Reconfiguration

Hybrid Health-Aware Supervisory Control Framework with a Prognostic Decision-Making

Jérôme Cieslak, David Gucik-Derigny, and Jing Chang

Abstract This paper deals with the design of a Hybrid Health-Aware Supervisory Control (HHASC) scheme for Linear Time Invariant (LTI) systems. The proposed HHASC setup integrates a switching fault-tolerant controller, a task of Fault Detection and Isolation (FDI), and the use of prognostic information. This approach is able to manage current and future health situations to extend the system life by the consideration of control objective mitigations to reduce the load on degraded components. In addition, the introduction of the Prognostic Decision-Making unit fitted with a Virtual Fault Mechanism allows to forecast the switching FTC and reduce the reconfiguration transients. The *dwell-time* property and its dual part are used to derive an input/output closed-loop stability proof. The proposed HHASC framework provides then a cohesive setup where prognostic, diagnosis and accommodation tasks can work in harmony together. The proposed solution is finally applied on a numerical aeronautical benchmark to highlight its benefit.

Keywords Fault-tolerant control · Prognostic decision-making · Supervisory control · Hybrid theory · Fault diagnosis

1 Introduction

In the health management domain, the research on Fault Detection and Isolation (FDI) has firstly received growing attention since the first patent in 1970. The concept of fault-tolerant computer emerged in 1970s has been progressively extended by the FDI community to introduce the Fault Tolerant Control (FTC) research field [1, 2]. FTC systems aim at guaranteeing that a process keeps fulfilling its mission even in

J. Cieslak (✉) · D. Gucik-Derigny
IMS Laboratory, University of Bordeaux, Bordeaux INP, CNRS (UMR 5218),
351 Cours de la Liberation, 33405 Talence, France
e-mail: jerome.cieslak@ims-bordeaux.fr

J. Chang
School of Aerospace Science and Technology, Xidian University, Xi'an, China

© The Author(s), under exclusive license to Springer Nature Switzerland AG 2021
J. Korbicz et al. (eds.), *Advances in Diagnostics of Processes and Systems*,
Studies in Systems, Decision and Control 313,
https://doi.org/10.1007/978-3-030-58964-6_1

the presence of failures, although possibly in a degraded mode. Even if there exist some solutions able to prove the closed-loop stability in spite of the inherent imperfect interactions between FDI and FTC parts [3–6], strategies work in reaction, i.e. after a delay. Hence, fault accommodation transients still exist and can unfortunately corrupt the mission. This makes the need of improved FTC strategies that can forecast a fault occurrence to engage the fault accommodation in a more appropriated instant, of great importance.

After a precursor work to introduce reliability in a FTC design [7], the Health-Aware Control (HAC) paradigm has been introduced in [8] with the emergence of prognostic schemes [9–11]. HAC helps to manage current and future health situations by updating the controller when the plant is still in healthy situations. The idea is to integrate prognostic information in the control policy [12, 13]. If some artificial intelligence tools [14–16] can be considered, the formalism of Model Predictive Control (MPC) has received growing attention [17–21]. This trend comes from the ease of introducing additional criteria (economic, ...) into the minimization of MPC cost function. Based on this setup, recent works propose some improvements to tackle the issues of Linear Parameter-Varying (LPV) plants [22, 23]. To use the efficient LPV design tools, LPV control architecture is proposed in [24]. This attempt uses a LPV filter to modify the motion control based on the Remaining Useful Lifetime (RUL). To overcome the possible conservativeness of LPV solutions, hybrid supervisory control have been introduced in [25, 26]. In all aforementioned solutions, there exists thus a mechanism based on prognostic information which is able to determine system actions: the Prognostic Decision-Making unit (PDM) [21].

In this paper, it is proposed to formalize the HAC problem by using hybrid theory. More precisely, the first contribution lies on the adaptation of the FTC architecture given in [5] to the paradigm of HAC. The proposed Hybrid Health-Aware Supervisory Control (HHASC) scheme implies the introduction of a Virtual Fault Mechanism (VFM) between the plant and switching control. The VFM and switching rules are next driven by the PDM to forecast the fault accommodation and reduce the load on degraded components. The second contribution is the establishment of the input/output closed-loop stability proof derived by using *dwell-time* and its dual part conditions. The proposed solution offers thus a unique context where a stability proof can be achieved to current (unforeseeable events leading to component damage) and future fault situations by the introduction of PDM unit. These enhancements permit thus to propose a reliable health management and mitigate the fault accommodation transients by forecasting a switch.

Notations A square matrix $X \in \Re^{n \times n}$ is Hurwitz if and only if all eigenvalues have a strictly negative real part. The symbol $|x|$ denotes its absolute value. For a matrix $X \in \Re^{n \times m}$, $|X| = \max_{1 \leq i \leq m} \sqrt{\lambda_i(X^T X)}$ denotes the induced norm where $\lambda_i(X^T X)$ is the ith eigenvalue of the matrix $(X^T X)$. For a function $d : \Re_+ \to \Re^d$, the L_∞ norm is $\|d\|_{[0,T)} = ess \sup_{0 \leq t \leq T} |d(t)|$, $\|d\| = \|d\|_{[0,+\infty)}$. All bounded signal d is noted L_∞^d. The symbol $\wedge$ corresponds to operator "and".

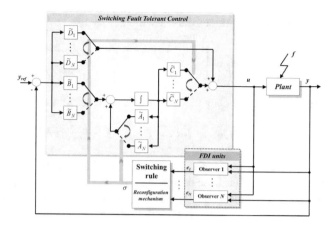

Fig. 1 The classical supervisory fault-tolerant control setup [5]

2 Problem Statement

Let the supervisory FTC setup of [5] be introduced, see Fig. 1. FDI task is achieved by a bank Luenberger-type observers fitted with a supervisor that drives a switching mechanism to select a controller. Let the disturbed healthy system be modeled by

$$\dot{x} = Ax + Bu + Gd, \quad y = Cx \tag{1}$$

where $x \in \mathfrak{R}^n$, $u \in \mathfrak{R}^m$, $y \in \mathfrak{R}^p$ and $d \in \mathfrak{R}^d$ are state, control, output and disturbances respectively.

Let the plant (1) be now formulated in hybrid manner to tackle actuator and component failure cases. These faults can be modeled as finite deviations of the nominal matrices A and B of (1). The faulty plant can thus be rewritten by

$$\dot{x} = A_i x + B_i u + G_i d, \quad y = Cx, \quad i = 1, \dots, N, \quad N > 1 \tag{2}$$

where $A_i = A + \Delta A_i$ and $B_i = B + \Delta B_i$. The system (2) defines a family of linear systems with the index $i \in I$ that can operate in N operating modes. The first mode ($i = 1$) denotes the fault-free situation and the $N - 1$ others, the faulty ones. To respect the pre-specified control specifications in each operating mode, let a bank of controllers K_i, $i = 1, \dots, N$ be considered. The state-space representations of K_i are

$$\dot{\tilde{x}}_i = \tilde{A}_i \tilde{x}_i + \tilde{B}_i (y_{ref} - y), \quad u = \tilde{C}_i \tilde{x}_i + \tilde{D}_i (y_{ref} - y) \tag{3}$$

with $i = 1, \dots, N$. $\tilde{x}_i \in \mathfrak{R}^{\tilde{n}_i}$ and $y_{ref} \in \mathfrak{R}^p$ are the state of (3) and reference signals respectively. Each controller has been designed to guarantee closed-loop stability and control performances when the index i is the same for the controller (3) and the plant (2), i.e. the following matrices are Hurwitz for $i = 1, \ldots, N$:

$$H_i = \begin{bmatrix} A_i - B_i \tilde{D}_i C & B_i \tilde{C}_i \\ -\tilde{B}_i C & \tilde{A}_i \end{bmatrix} \tag{4}$$

The Luenberger observers obey to this state-space representation

$$\dot{z}_i = A_i z_i + B_i u + L_i (y - C z_i), \quad i = 1, \ldots, N \tag{5}$$

where $z_i \in \Re^n$ is a state estimate of (2) and L_i denotes the ith observer gain. Obviously, the observer gain L_i has been designed such that $A_i - L_i C$ is Hurwitz for all $i = 1, \ldots, N$. When the indexes of (5) and (2) are the same, the estimation error $e_i = x - z_i$ is given by $\dot{e}_i = (A_i - L_i C) e_i + G_i d$. Under bounded disturbance d, estimation error is thus asymptotically stable. The supervisor uses this property to identify the current operating mode, i.e. the estimator that yields the smallest error allows the identification of the current working mode. The switching logic is thus a decision map $\mathcal{H} : \Re^m \times \Re^p \times \Re^{n \times N} \to \mathcal{I}$ that generates the piecewise constant signal

$$\sigma(t) = \mathcal{H}(u, y, z_1, \ldots, z_N) \tag{6}$$

to engage the corresponding controller (3) to control the plant (2). Here, a hysteresis-based switching logic is used, i.e. it follows for $k \geq 0$ and $t_0 = 0$

$$t_{k+1} = \underset{t \geq t_k + \tau_D}{\arg\inf} \left\{ t : \left| C e_{\sigma(t_k)}(t) \right| > h \left| C e_j(t) \right|, j = 1, \ldots, N, j \neq \sigma(t_k) \right\} \tag{7}$$

$$\sigma(t_k) = \underset{1 \leq j \leq N}{\arg\min} \left| C e_j(t_k) \right|, k \geq 0 \tag{8}$$

$$\sigma(t) = \sigma(t_k) \; for \; all \; t_k \leq t < t_{k+1}, k \geq 0 \tag{9}$$

where $t_k, \; k \geq 0$ are switching instants and the *dwell-time* constant is $\tau_D > 0$. $e_{\sigma(t_k)}(t)$ is to the estimation error $x - z_{\sigma(t_k)}$ where $\sigma(t_k)$ denotes the index of engaged controller. At $t_0 = 0$, $\sigma(t)$ is initialized as $\sigma(t_0) = 1$ and the rule (8) is valid for other time instants t_k. The new switching time t_{k+1} is driven by (7), i.e. t_{k+1} is generated if and only if $\left| C e_{\sigma(t_k)}(t) \right| > h \left| C e_j(t) \right|$. The parameter $h \geq 1$ defines the hysteresis between switches to prevent the risk of chattering due to the disturbances [27]. According to [5], the setup of Fig. 1 obeys to

$$\dot{\varsigma}_k = W_{k,i} \varsigma_k + V_{k,i} C e_k + \tilde{G}_i d + \bar{G}_i y_{ref} \tag{10}$$

where $\varsigma_k = [z_k^T \tilde{x}_k^T x^T z_1^T \cdots z_N^T]^T$ such that $z_1^T \cdots z_N^T$ part does not contained z_k^T (for simplicity we use shorthand notation $\sigma(t_k) = k$). The matrix $W_{k,i}$ is a lower triangular matrix where its blocks in the main diagonal are $H_k, A_i - L_i C, A_1 -$

$L_1 C, \ldots, A_N - L_N C$. Since all blocks are Hurwitz, the exponential stability with switching logic (7)–(9) can be guaranteed by the following theorem.

Theorem 1 ([5]) *Let* $A_i - L_i C$ *and* (4) *be Hurwitz for* $i = 1, \ldots, N$. *For a plant index* $i(t) = cst$ *and all* $t \geq 0$, *there exists* $\tau_D > 0$ *such that the hybrid system composed of* (2), (3), (5) *and* (6) *possesses the property*

$$|\varsigma(t)| \leq v_i e^{-\mu_i t / \tau_D} |\varsigma(0)| + v_i \left\{ \|\delta\|_{[0,t)} + \|d\|_{[0,t)} + \|y_{ref}\|_{[0,t)} \right\} \qquad (11)$$

for any $\varsigma(0) \in \mathfrak{R}^{(N+1)n + \tilde{n}_k}$, $\varsigma = [z_k^T \tilde{x}_k^T x^T z_1^T \cdots z_N^T]^T$, $d \in L_\infty^d$, $y_{ref} \in L_\infty^{y_{ref}}$, $v_i > 0$, $\mu_i > 0$, $v_i > 0$ *and where*

$$\delta(t) = \begin{cases} C[e_i(t) - e_{\sigma(t_k)}(t)] \text{ if } t \in [t_k, t_k + \tau_D) \wedge |Ce_i(t)| < |Ce_{\sigma(t_k)}(t)| \\ 0 \text{ otherwise} \end{cases}.$$

Based on Theorem 1, exponential stability is achieved for τ_D given by [28]

$$\tau_D = \max_{1 \leq j \leq N} -\alpha_{j,i}^{-1} \ln(\lambda \beta_{j,i}^{-1}) \qquad (12)$$

that denotes the highest authorized time interval between two switches for controllers (3). $0 < \lambda < 1$ is a design constant and $\alpha_{j,i}$ is the minimal eigenvalue of the norm real part of $W_{j,i}$. $\beta_{j,i} = \sup_{t \geq 0} |\exp(W_{j,i}t)|$ for $j = 1, \ldots, N$, $j \neq i$. If the assumption $i(t) = cst$ for all $t \geq 0$ is relaxed, the dual *dwell-time* T_D takes place for stability proof. T_D can be seen as the minimum admissible time interval between two consecutive faults to guarantee the switched system stability, i.e.:

Corollary 1 ([5]) *Let* $A_i - L_i C$ *and* (4) *for* $i = 1, \ldots, N$ *be Hurwitz. Suppose that the plant index* $i(t) = i(T_r)$ *during a time interval* T_r, *i.e.* $t \in [T_r, T_{r+1})$, $T_{r+1} - T_r \geq T_D$ *and* $i(T_r) \in \mathcal{I}$ *for all* $r \geq 0$. *There exists* $T_D > 0$ *and* $\tau_D > 0$ *such that the system* (2), (3), (5) *and* (6) *possesses the property*

$$|\varsigma(t)| \leq \tilde{v} e^{-\tilde{\mu} t / T_D} |\varsigma(0)| + \tilde{v} \left\{ \|\delta\|_{[0,t)} + \|d\|_{[0,t)} + \|y_{ref}\|_{[0,t)} \right\} \qquad (13)$$

for any $\varsigma(0) \in \mathfrak{R}^{(N+1)n + \tilde{n}_k}$, $\varsigma = [z_k^T \tilde{x}_k^T x^T z_1^T \cdots z_N^T]^T$, $d \in L_\infty^d$, $y_{ref} \in L_\infty^{y_{ref}}$, $\tilde{v} > 0$, $\tilde{\mu} > 0$, $\tilde{\xi} > 0$ *and* $\tilde{v} > 0$.

From Corollary 1, (13) is bounded on interval $[T_r, T_{r+1})$, $r \geq 0$ if T_D is [5]

$$T_D = \max_{1 \leq j \leq N} \left\{ -\tau_D \mu_j^{-1} \ln(\bar{\lambda} v_j^{-1}) \right\} \qquad (14)$$

where $0 < \bar{\lambda} < 1$ is a constant design parameter. The other variables are given by $\beta_{j,i} = \sup_{t \geq 0} |\exp(W_{j,i}t)|$, $\mu_j = -0.25 \ln(\lambda)$, $\gamma_{j,i} = |W_{j,i}^{-1} V_{j,i}|$, $\bar{\gamma}_i = \max_{1 \leq j \leq N} \gamma_{i,j}$, $v_j = \bar{\beta}_j e + (e - 4 \ln(\lambda)^{-1} \bar{\beta}_j \lambda^{-1}) \bar{\gamma}_j \rho_1$ and $\bar{\beta}_i = \max_{1 \leq j \leq N} \beta_{i,j}$, see [5] for details.

In the following, the HHASC setup is introduced, see Fig. 2. HHASC will use the *dwell-time* τ_D and its dual part T_D in a modified supervisor and a PDM unit. FDI task is still achieved by observers (5). Switching controllers obey to (3). Switching logic introduced in (7)–(9) is maintained with a slight change defined in Sect. 3. The main HHASC feature lies on its capability to forecast the future need of switching thanks to a VFM driven by the PDM.

3   Main Result

Let the changes in the supervisor be firstly discussed. In parallel to (7)–(9), the modified supervisor has a monitoring of changes on new input $\sigma_P(t)$ provided by PDM. A condition $Edge\{\sigma_P(t)\}$ has thus been added, see Algorithm 1. $Edge\{\bullet\}$ denotes a rising or falling edge of a signal. The use of this condition on the HHASC allows the enforcement of switches driven by the PDM unit and, *de facto*, the reduction of bump effects on signals [4, 6] due to this forecasting action.

Let the PDM algorithm be now introduced. Its goal is to generate a signal $\sigma_P(t)$ to the modified supervisor and able to disengage the future fault—outage failure case—by acting on the VFM. Here, inputs of PDM are the remaining useful life time $T_{RUL_i}(t), i = 1, \ldots, N_P, N_P \leq N$ of the N_P components monitored by a prognosis scheme. The condition $N_P \leq N$ has just been proposed to take into account that it can be costly and not always possible to implement a prognostic solution in practical engineering applications [29]. The PDM obeys to the following rules

$$t_{r+1}^P = \underset{t \geq t_r^P + T_D, \quad j \in (\mathcal{N}^* - \{\sigma_P(t_r^P)\})}{\arg\inf} \left\{ t : \left| T_{rul_j}(t) \right| < \eta T_D \right\} \tag{15}$$

$$\sigma_P(t_r^P) = \underset{j \in (\mathcal{N}^* - \{\sigma_P(t_r^P)\})}{\arg\min} \left| T_{rul_j}(t_r^P) \right| \tag{16}$$

$$\sigma_P(t) = \sigma_P(t_r^P) \; for \; all \; t_r^P \leq t < t_{r+1}^P, r \geq 0 \tag{17}$$

where t_r^P, $r \geq 0$ are switching instants and T_D is the dual *dwell-time*. At $t_0^P = 0$, σ_P is initialized as $\sigma_P(t_0^P) = \{1\}$ and the rule (16) is valid for all time instants t_r. The new switching time t_{r+1}^P is driven by (15), i.e. the instant t_{r+1}^P exists if and only if $\left| T_{rul_j}(t) \right| < \eta T_D$. Note that $\mathcal{N}^* = \{2, \ldots, N_P\}$ is the set of N_P possible RUL and that it is proposed to discard the already identified outages. $\eta \geq 1$ defines a tuning parameter.

Remark 1   Note that other prognostic indicators (degradation index, ...) [29] can be considered by HHASC setup. Since the generation of a reliable $T_{RUL_i}(t)$ is not the paper objective, it is also assumed that there is a high confidence level of $T_{RUL_i}(t)$.

From Theorem 1 and Corollary 1, stability with the modified supervisor and the PDM (15)–(17) given in Algorithm 1 is achieved by the next theorem.

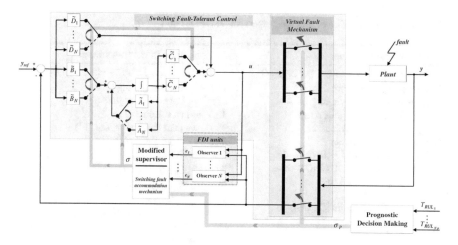

Fig. 2 The proposed HHASC architecture

Theorem 2 *Let $A_i - L_i C$ and (4) for $i = 1, \ldots, N$ be Hurwitz. Suppose that $i(t) = i(T_r), t \in [T_r, T_{r+1}), T_{r+1} - T_r \geq T_D$ and $i(T_r) \in \mathcal{I}$ for all $r \geq 0$. There exists $T_D > 0$ and $\tau_D > 0$ such that hybrid system composed of (2), (3), (5) and (6), the modified supervisor and PDM (15)–(17) given in Algorithm 1 obeys to*

$$|\varsigma(t)| \leq \tilde{v} e^{-\tilde{\mu}t/T_D} |\varsigma(0)| + \tilde{v} \left\{ \|\delta\|_{[0,t)} + \|d\|_{[0,t)} + \|y_{ref}\|_{[0,t)} \right\} \quad (18)$$

for any $\varsigma(0) \in \mathfrak{R}^{(N+1)n + \tilde{n}_k}$, $\varsigma = [z_k^T \ \tilde{x}_k^T \ x^T \ z_1^T \ \cdots \ z_N^T \ ]^T$, $d \in L_\infty^d$, $y_{ref} \in L_\infty^{y_{ref}}$, $\tilde{v} > 0$, $\tilde{\mu} > 0$, $\tilde{v} > 0$ and where

$$\delta(t) = \begin{cases} C[e_i(t) - e_{\sigma(t_k)}(t)] \ if \ t \in [t_k, t_k + \tau_D) \wedge |Ce_i(t)| < |Ce_{\sigma(t_k)}(t)| \\ 0 \ otherwise \end{cases}.$$

Proof The hybrid system (2), (3), (5), (6) equipped with the modified supervisor and the PDM unit described in Algorithm 1, is a linear and continuous system on intervals $[t_k, t_{k+1}), k \geq 0$, i.e. signals are continuous for all $t \geq 0$ and $d \in L_\infty^d, y_{ref} \in L_\infty^{y_{ref}}$. Based on *dwell-time* conditions, finite bumps can appear after a switch but the continuity is not destroyed. So, signal $|Ce_{\sigma(t_k)}(t)|$ is bounded on $[t_k, t_{k+1}), k \geq 0$. Since $|Ce_{\sigma(t_k)}(t)| \leq |Ce_i(t)| + |\delta(t)|$ for all $t \geq 0$ with the definition of δ, the following upper estimate is true for $d \in L_\infty^d$ and $y_{ref} \in L_\infty^{y_{ref}}$:

$$|Ce_{\sigma(t_k)}(t)| \leq \rho_1 |e_i(0)| e^{-\eta t} + \rho_2 \|d\| + \rho_3 \|y_{ref}\| + \|\delta\| \quad (19)$$

In spite of the presence of VFM and PDM unit, the HHASC setup of Fig. 2 obeys to the same state-space representation that the one given in (10), i.e. all diagonal blocks are Hurwitz. Since the VFM switching is now synchronized with the switching of supervisory fault-tolerant control, the modified supervisor algorithm stops after a

Algorithm 1 Description of modified supervisor and PDM unit

Modified supervisor:	Prognostic Decision-Making:
Inputs: $\sigma_P, e_i, i \in \mathcal{M} = \{1, \ldots, N\}$	**Inputs:** $T_{rul_j}, j \in \mathcal{N}^* = \{2, \ldots, N_P\}$
1: $k \leftarrow$ current sampling time	1: $k \leftarrow$ current sampling time
2: $t_k \leftarrow$ last control switching $(t_k < k)$	2: $t_r^P \leftarrow$ last switching time $(t_r^P < k)$
3: $l_s \leftarrow$ last selected index $\sigma(t_k)$	3: $l_s \leftarrow$ last selected index $\sigma_P(t_r^P)$
4: $\tau_D \leftarrow$ dwell-time value, see (12)	4: $T_D \leftarrow$ dual *dwell-time*, see (14)
5: $h \leftarrow$ design parameter	5: $\eta \leftarrow$ design parameter
6: **for each** $e_i, i \in (\mathcal{M} - \{\sigma(t_k)\})$ **do**	6: **for each** $T_{rul_j}, j \in (\mathcal{N}^* - \{\sigma_P(t_r^P)\})$ **do**
7: **if** $\left|Ce_{\sigma(t_k)}(t)\right| > h \left|Ce_i(t)\right|$ **then**	7: **if** $\left|T_{rul_j}(k)\right| < \eta T_D$ **then**
8: **if** $k \geq t_k + \tau_D$ **then**	8: **if** $k \geq t_r^P + T_D$ **then**
9: $t_k = k$;	9: $t_r^P = k$;
10: $l_s = \arg\min |Ce_i(k)|$;	10: $l_s = \arg\min \left|T_{rul_j}(k)\right|$;
11: **end if**	11: **end if**
12: **end if**	12: **end if**
13: **end for**	13: **end for**
14: **for** σ_P **do**	14: $\sigma_P(k) = l_s$;
15: **if** *Edge* $\{\sigma_P(k)\}$ **then**	
16: **if** $k \geq t_k + \tau_D$ **then**	
17: $t_k = k$;	
18: $l_s = \sigma_P(k)$;	
19: **end if**	
20: **end if**	
21: **end for**	
22: $\sigma(k) = l_s$;	

finite number of switches $k \geq 0$. Hence, observers (5) allow the identification $\sigma(t_k)$ with the single minimum among the observation errors e_j, $j = 1, \ldots, N$, i.e. if the distinguishability property is valid [30].

Remark 2 Luenberger observers can be limited to always achieve the distinguishability property. Based on [30], it can be noticed that the main result holds to any type of linear diagnosis filters.

To achieve exponential stability, the following upper bound for the state of (10) can be considered for $t \in [t_k, t_k + 1)$:

$$|\varsigma(t)| \leq \beta_{k,i} |\varsigma(t_k)| e^{-\alpha_{k,i}(t-t_k)} + \gamma_{k,i} \left\|Ce_{k,i}\right\|_{[t_k,t)} + \chi_{k,i} \|d\| + \varpi_{k,i} \left\|y_{ref}\right\|$$

$$\left\|Ce_k\right\|_{[t_k,t)} = ess \sup_{t_k \leq \tau < t} |Ce_k(\tau)| \leq \rho_1 |e_i(0)| e^{-\pi t_k} + \rho_2 \|d\| + \rho_3 \left\|y_{ref}\right\| + \|\delta\|_{[t_k,t)}$$

where the shorthand notation $k = \sigma(t_k)$ has been used for the signal indexes. $\alpha_{j,i}$ is still the minimal eigenvalue in norm real part of the $W_{j,i}$, $\chi_{j,i} = \left|W_{j,i}^{-1} \tilde{G}_i\right|$, $\gamma_{j,i} = \left|W_{j,i}^{-1} V_{j,i}\right|$, $\beta_{j,i} = \sup_{t \geq 0} \left|\exp(W_{j,i}t)\right|$, and $\varpi_{j,i} = \left|W_{j,i}^{-1} \bar{G}_i\right|$ for $j = 1, \ldots, N, j \neq i$. By previous definitions, $|e_i(0)| \leq |\varsigma(0)|$ and $\alpha_{j,i} \leq \pi$ for all $j = 1, \ldots, N$. From

[5, 28], τ_D can be obtained by using (12) where $0 < \lambda < 1$ is a constant defined such that $e^{-\pi t_k} \leq \lambda^k$, $k \geq 0$. The boundedness of the hybrid system state of Fig. 2 gives for all $t \geq 0$:

$$|\varsigma(t)| \leq e\left(\bar{\beta}_i(1 - 4\ln(\lambda)^{-1}e^{-1}\bar{\gamma}_i\rho_1\lambda^{-1}) + \bar{\gamma}_i\rho_1\right)e^{0.25\ln(\lambda)t\tau_D^{-1}}|\varsigma(0)|$$
$$+ \left(1 + \bar{\beta}_i(1 - \lambda)^{-1}\right)\left(\bar{\gamma}_i\|\delta\| + \bar{\chi}_i\|d\| + \bar{\varpi}_i\|y_{ref}\|\right) \quad (20)$$

From (11) and (20), it is possible to identify the expression $\mu_i = -0.25\ln(\lambda)$, $v_i = \bar{\beta}_i e + (e - 4\ln(\lambda)^{-1}\bar{\beta}_i\lambda^{-1})\bar{\gamma}_i\rho_1$ and $v_i = \max\{\bar{\gamma}_i, \bar{\chi}_i, \bar{\varpi}_i\}(1 + \bar{\beta}_i(1 - \lambda)^{-1})$. Let $i(t) = i(T_r)$, $t \in [T_r, T_{r+1})$ and $i(T_r) \in \mathcal{I}$ for all $r \geq 0$ be now introduced. The controlled plant (2) can thus be different on interval $[T_r, T_{r+1})$ and $i(t)$ is a piecewise constant signal identifying a change of real or virtual (due to VFM) operating mode. Since this situation has been considered in the design step, there exists at least one stabilizing controllers (3) and one Hurwitz observer (5) to manage real or virtual modes. It follows that all conditions of theorem 2 hold on any intervals $[T_r, T_{r+1})$, $r \geq 0$. The upper estimate of $\varsigma(t)$ (18) can thus be defined by

$$|\varsigma(t)| \leq v_r e^{-\mu_r(t-T_r)/\tau_D}|\varsigma(T_r)| + v_r\left\{\|\delta\|_{[T_r,t)} + \|d\|_{[T_r,t)} + \|y_{ref}\|_{[T_r,t)}\right\} \quad (21)$$

where τ_D, is obtained by (12) and v_r, μ_r, and v_r can be derived from expressions after (20). Using the developments of [5], it can be shown that the consideration of the dual *dwell-time* (14) leads to the following result

$$|\varsigma(t)| \leq e\bar{v}e^{0.5\ln(\bar{\lambda})t/T_D}|\varsigma(0)| + \tilde{v}\left\{\|\delta\| + \|d\| + \|y_{ref}\|\right\} \quad (22)$$

with $\bar{v} = \max_{1 \leq j \leq N} v_j$, $\tilde{v} = \bar{v}(1 + \bar{v}(1 - \lambda)^{-1})$ and $\bar{v} = \max_{1 \leq j \leq N} v_j$. Choosing $\tilde{v}$ and $\tilde{\mu}$ according to $\tilde{v} = e\bar{v}$ and $\tilde{\mu} = 0.5\ln(\bar{\lambda})$ respectively, it follows that equation (22) is equivalent to (18). This completes the proof.

4 Numerical Example

An application to the longitudinal model of the highly maneuverable technology (*HiMAT*) vehicle is considered [31]. The model possesses three inputs: the elevators δ_s, elevons δ_e and canard flaps δ_c. The vehicle is equipped with two sensors measuring the angle of attack y_α and the pitch rate y_q given in rad and rad/s respectively. The trajectory of flight concerns a vertical translation, i.e. the pitch rate y_q is null whereas the attack angle y_α changes. The plant can be modeled as follows

$$\dot{x}(t) = A_i x(t) + B_i u(t), i = 1, 2, 3 \quad (23)$$

with $u = (\delta_s, \delta_e, \delta_c)^T$ and the plant state x corresponds to the attack angle and the pitch rate. Matrices A_i and B_i with $i = 1, 2, 3$ are state-space representations of healthy and fault modes. The index $i = 1$ is dedicated to healthy system where:

$$A_1 = \begin{bmatrix} -1.0772 & 0.96528 \\ 9.068 & -1.5077 \end{bmatrix}, \quad B_1 = \begin{bmatrix} -0.17211 & -0.12245 & -0.01431 \\ -7.9948 & -4.955 & 5.0369 \end{bmatrix}$$

In this study case, two faulty situations are investigated. The index $i = 2$ denotes a situation where the first actuator (monitored by a prognosis scheme that generates a reliable T_{RUL} estimate) is out of order. Last fault situation ($i = 3$) considers the complete outage of first actuator and a Loss of Effectiveness (LoE) of 50% of the second actuator. Since these faulty situations are recoverable, three control laws (3) with integral action are designed using a linear quadric approach [6]. Note that controllers with index 2 and 3 drive the *HiMAT* vehicle by using only the second and third actuator, i.e. there is a very weak structural load on the first actuator leading to an extension of aircraft life in spite of this outage. Let three Luenberger observers (5) be now designed such that $A_i - L_i C$ holds. The observer poles $i = 1, 2, 3$ are arbitrarily fixed to $[-16, -15], [-56, -55]$ and $[-6, -5]$ respectively (see [5] for an optimization). τ_D and T_D has been computed by using (12) and (14). It follows that $\tau_D = 1.5793$ s and $T_D = 21.3207$ s. Note that results are exactly the same than those obtained with [5]. The architectures depicted in Figs. 1 and 2 are now implemented within the *HiMAT* simulator in Matlab/Simulink environment. The elevator outage appears at $t = 28$ s and an intermittent failure case has been considered during the time intervals [75, 100] s and [107, 120] s. The hysteresis parameter h involved in (7) has been fixed to 50. The parameter η is fixed to 1.5.

To emphasize the benefit of the HHASC scheme, same simulations are performed with the classical supervisory FTC setup [5], with the proposed solution without the modified supervisor (i.e. without the lines 14–21 of the Algorithm 1) and with the proposed HHASC setup. As it can be seen in Fig. 3, all solutions maintain stability and keep acceptable performance level in terms of tracking specifications (null static error). Around 14 s, the T_{RUL} is under the threshold ηT_D. A switch can thus be triggered. Since it is necessary to wait a delay of T_D between two virtual or real hybrid plant configurations, the forecasted switch only comes at 21.32 s for the HHASC. On the pitch rate and attack angle signals of Fig. 3, cyan curves show the result when the modified supervisor is not considered. Without the expected synchronization between the VFM and the switching control algorithm, there are bumps around 21 s since it is necessary to wait the detection of the virtual fault by the bank of observers. Fortunately, this weakness disappears with the proposed HHASC. In addition, there is a clear change on the behavior of the T_{RUL} due to a mitigation of structural load on the first actuator outage. The explanation of such behavior change comes from a control signal of the first actuator equal to zero after forecasted switch, see the bottom plots of Fig. 4. From 75 s, it can be seen that the proposed HHASC scheme possesses similar performances than [5]. This expected result comes from the need to detect a fault occurrence by using observers (5) and the switching rules (7)–(9) to engage a fault accommodation. This facet confirms that prognosis and diagnosis tasks can work in harmony together. Finally, the interval between two different plant configurations is equal to 7 s during the intermittent failure case, see after 75 s. Since

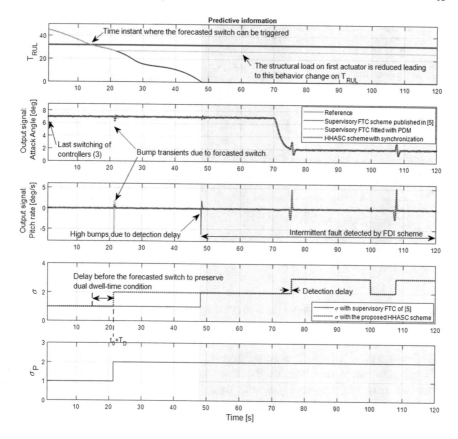

Fig. 3 Simulation results for three different FTC schemes

this value is smaller than T_D, stability cannot be theoretically guaranteed. However, it can be seen that the system stays stable. This point underlines that τ_D and T_D could be, at this stage, a conservative solution.

5 Conclusion

The problem of designing a Hybrid Health-Aware Supervisory Control (HHASC) scheme for Linear Time Invariant (LTI) systems is addressed. The main contribution of the proposed HHASC setup is the introduction of a Prognostic Decision-Making unit and a Virtual Fault Mechanism in the supervisory FTC setup to allow the management of current and future health situations. By this way, it is possible to forecast the switch of control algorithm to reconfiguration transient mitigation purpose and plant life extension by reducing the structural load on degraded components. In

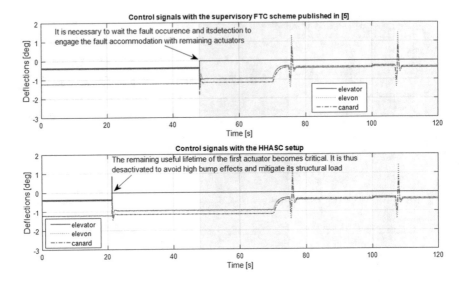

Fig. 4 Control signals with solution [5] (top) and the proposed HHASC (bottom)

addition, a formal input/output stability proof is proposed by using the properties of *dwell-time* and its dual part. Further investigations are necessary to integrate the stochastic nature of prognostic information in the HHASC scheme. Based on [32], it is the topic of our current research.

References

1. Yu, X., Jiang, J.: A survey of fault-tolerant controllers based on safety-related issues. Ann. Rev. Control **39**, 46–57 (2015)
2. Zolghadri, A., Henry, D., Cieslak, J., Efimov, D., Goupil, P.: Fault Diagnosis and Fault-Tolerant Control and Guidance for Aerospace Vehicles: From Theory to Application. Advances in Industrial Control. Springer, London (2014)
3. Guo, Z., Zhou, J., Guo, J., et al.: Coupling-characterization-based robust attitude control scheme for hypersonic vehicles. IEEE Trans. Ind. Electron. **64**(8) (2017)
4. Jain, T., Yamé, J.J., Sauter, D.: Model-free reconfiguration mechanism for fault tolerance. Int. J. Appl. Math. Comput. Sci. **22**(1), 125–137 (2012)
5. Efimov, D., Cieslak, J., Henry, D.: Supervisory fault-tolerant control with mutual performance optimization. Int. J. Adapt. Control Signal Proc. **27**, 251–279 (2013)
6. Cieslak, J., Efimov, D., Henry, D.: Transient management of a supervisory fault-tolerant control scheme based on dwell-time conditions. Int. J. Adapt. Control Signal Proc. **29** (2015)
7. Guenab, F., Theilliol, D., Weber, P., Zhang, Y.M., Sauter, D.: Fault tolerant control system design: a reconfiguration strategy based on reliability analysis under dynamic behavior constraints. IFAC Proc. **39**(13), 1312–1317 (2006)
8. Escobet, T., Puig, V., Nejjari, F.: Health aware control and model-based prognosis. In: 20th IEEE Mediterranean Conference on Control and Automation (2012)

9. Tang, L., Kacprzynski, G.J., Goebel, K., Vachtsevanos, G.: Case studies for prognostics-enhanced automated contingency management for aircraft systems. In: IEEE Aerospace Conference, Big Sky, USA (2010)

10. Gucik-Derigny, D., Outbib, R., Ouladsine, M.: A comparative study of unknown-input observers for prognosis applied to an electromechanical system. IEEE Trans. Reliab. **65**(2), 704–717 (2016)

11. Yousfi, B., Raïssi, T., Amairi, M., Aoun, M.: Set-membership methodology for model-based prognosis. ISA Trans. **66**, 216–225 (2016)

12. Traore, M., Chammas, A., Duviella, E.: Supervision and prognosis architecture based on dynamical classification method for the predictive maintenance of dynamical evolving systems. Reliab. Eng. Syst. Saf. (2015)

13. Langeron, Y., Grall, A., Barros, A.: Joint maintenance and controller reconfiguration policy for a gradually deteriorating control system. Proc. Inst. Mech. Part O: J. Risk Reliab. **231**(4), 339–349 (2017)

14. Choo, B.Y., Adams, S., Beling, P.: Health-aware hierarchical control for smart manufacturing using reinforcement learning. In: International Conference on Prognostics and Health Management, PHM (2017)

15. Chatila, R., Renaudo, E., Andries, M., et al.: Toward self-aware robots. Front. Robot. AI **5**, 88 (2018)

16. Jha, M.S., Weber, P., Theilliol, D., Ponsart, J.C., Maquin, D.: A reinforcement learning approach to health aware control strategy. In: 27th IEEE Mediterranean Conference on Control and Automation, pp. 171–176 (2019)

17. Verheyleweghen, A., Jäschke, J.: Framework for combined diagnostics, prognostics and optimal operation of a subsea gas compression system. IFAC-PapersOnLine **50**(1), 15916–15921 (2017)

18. Pour, F.K., Puig, V., Ocampo-Martinez, C.: Multi-layer health-aware economic predictive control of a pasteurization pilot plant. Int. J. Appl. Math. Comput. Sci. 28(1), 97–110 (2018)

19. Wang, Y., De la Peña, D.M., Puig, V., Cembrano, G.: Robust economic model predictive control based on a periodicity constraint. Int. J. Robust Nonlinear Control **29**(11), 3296–3310 (2019)

20. Mrugalska, B., Stetter, R.: Health-aware model-predictive control of a cooperative AGV-based production system. Sensors **19**, 532 (2019)

21. Balaban, E., Alonso, J.J., Goebel, K.F.: An approach to prognostic decision making in the aerospace domain. In: Annual Conference of the Prognostics and Health Management Society (2012)

22. Pour, F.K., Puig, V., Cembrano, G.: Economic health-aware LPV-MPC based on system reliability assessment for water transport network. Energies (2019)

23. Pour, F.K., Puig, V., Cembrano, G.: Economic reliability-aware MPC-LPV for operational management of flow-based water networks including chance-constraints programming. Processes **8**, 60 (2020)

24. Rodriguez, D.J., Martinez, J.J., Bérenguer, C.: An architecture for controlling the remaining useful lifetime of a friction drive system. IFAC-PapersOnLine **51**(24), 861–866 (2018)

25. Cieslak, J., Gucik-Derigny, D.: Introduction of a prognostic decision making in a fault tolerant control context. IFAC-PapersOnLine **51**(24), 649–654 (2018)

26. Gross, K.C., Baclawski, K., et al.: A supervisory control loop with prognostics for human-in-the-loop decision support and control applications. In: IEEE Conference on Cognitive and Computational, Aspects of Situation Management (2017)

27. Hespanha, J.P., Liberzon, D., Morse, A.S.: Overcoming the limitations of adaptive control by means of logic-based switching. Syst. Control Lett. **49**, 49–65 (2003)

28. Liberzon, D.: Switching in Systems and Control. Birkhäuser, Boston (2003)

29. Huynh, K.T., Grall, A., Bérenguer, C.: A parametric predictive maintenance decision-making framework considering improved system health prognosis precision. IEEE Trans. Reliab. **68**(1), 375–396 (2019)

30. Cieslak, J., Henry, D., Efimov, D., Zolghadri, A.: Enhanced Distinguishability in Supervisory Fault Tolerant Control. IFAC World Congress, Cape Town (2014)

31. Hou, Y., Dong, C., Wang, Q.: Stability analysis of switched linear systems with locally over-
    lapped switching law. J. Guid. Control Dyn. **33**(2), 396–403 (2010)
32. Demirel, B., Briat, C., Johansson, M.: Deterministic and stochastic approaches to supervisory
    control design for networked systems with time-varying communication delays. Nonlinear
    Anal. Hybrid Syst. **10**, 94–110 (2013)

Reconfiguration of Nonlinear Faulty Systems via Linear Methods

Alexey Zhirabok, Evgeny Bobko, and Artur Filatov

Abstract The problem of reconfiguration in faulty systems containing non-smooth nonlinearities is considered. To solve the problem, the control law is constructed providing full decoupling with respect to fault effects. The suggested solution is based on so-called logic-dynamic approach the main feature of which is that it allows to use linear methods for systems with non-smooth nonlinearities. Calculating relations are given for the control law. Theoretical results are illustrated by example.

Keywords Nonlinear systems · Faults · Reconfiguration

1 Introduction

In this paper, the problem of fault tolerant control (FTC) in technical systems of critical purposes is studied. There are two main approaches to the FTC. The first one is fault accommodation while the second one is system reconfiguration [2, 8–11]. This paper concentrates on reconfiguration in technical systems described by nonlinear dynamic models. In our approach, we interpret the faults as unknown disturbances and use the known solution of the disturbance decoupling problem (DDP) to solve the system reconfiguration problem (SRP).

Two solutions of SRP have been proposed. The first one uses the methods of adaptive control and assumes that faults are detected and estimated and then the control law accommodation is designed [2, 3, 11]. The second approach assumes that the faults are already detected and isolated and a control law is determined. The

A. Zhirabok (✉) · E. Bobko
Far Eastern Federal University, Vladivostok 690091, Russia
e-mail: zhirabok@mail.ru

E. Bobko
e-mail: evgbobko@gmail.com

A. Zhirabok · A. Filatov
Institute of Marine Technology Problems, Vladivostok 690091, Russia
e-mail: mygryphon@gmail.com

© The Author(s), under exclusive license to Springer Nature Switzerland AG 2021
J. Korbicz et al. (eds.), *Advances in Diagnostics of Processes and Systems*,
Studies in Systems, Decision and Control 313,
https://doi.org/10.1007/978-3-030-58964-6_2

law is such that some function of the system output is full decoupled with respect to effects of faults [7]. Unlike the first approach, the fault estimation is not required in the second approach.

The problem of reconfiguration in nonlinear dynamic systems with non-smooth nonlinearities was considered in [5] and then in [13, 14]. The peculiarity of [5] is that the known solution of the DDP is used to solve the PRP. Besides, a solution is based on the functions' algebra and demand complex analytical calculations. In contrast to [5], the papers [13, 14] use so-called logic-dynamic approach developed in [15]; a solution is not based on the DDP. The goal of the present paper is to find a solution of the SRP by the methods of linear algebra only without nonlinear calculations using the logic-dynamic approach; a solution will be based on the DDP.

To achieve this goal, the mail results of the functions' algebra based solution of the DDP are presented at first. Next this problem is solved by linear methods based on the logic-dynamic approach. Finally, the obtained solution is transformed to find a solution of the SRP.

2 Preliminaries

The DDP was solved in [4] for system described by general nonlinear model

$$x(t + 1) = f(x(t), u(t), d(t)),$$
$$y(t) = h(x(t)), \tag{1}$$

where $x \in R^n$, $u \in R^m$, $y \in R^l$ are vectors of state, control and output; f and h are arbitrary nonlinear functions; $d(t) \in R^p$ is the function presenting faults: if faults are absent, $d(t) = 0$, when a fault occurs, $d(t)$ becomes an unknown function of time. Similar to [5], in the present paper the faults are interpreted as disturbances and the solution of the DDP is used to solve the PRP.

The DDP is stated as follows: given the function $y_*(t) = h_*(x(t))$, construct the compensator S_0 described by the model

$$x_0(t + 1) = f_0(x_0(t), u(t), y(t)),$$
$$u(t) = g_0(x_0(t), u_*(t), y(t)), \tag{2}$$

so that the output $y_*(t)$, $t \geq 0$, in the close-looped system is independent from the function $d(t)$, where $u_*(t)$ is a new control.

Recall some definitions and results from [4, 5]. The vector function α is said to be (h, f)-invariant if

$$\alpha(f(x, u, d)) = f_*(\alpha(x), h(x), u, d)$$

for some function f_* and all x, u, and d. The vector function α is said to be f-invariant if

$$\alpha(f(x, u, d)) = f_*(\alpha(x), u, d)$$

for some function f_* and all x, u, and d. The vector function ξ is said to be a controlled invariant if a regular static state feedback $u = g'(x, u_*)$ exists such that the function ξ is f-invariant in the closed-loop system (1), (2).

Theorem 1 ([5]) *The output-to-be-controlled y_* of system (1) can be disturbance decoupled by compensator (2) iff (h, f)-invariant function α and a controlled invariant function ξ exist such that*

$$\alpha^0 \le \alpha \le \xi \le h_*, \tag{3}$$

where the inequality $\beta \le \gamma$ means that the function δ exists such that $\delta(\beta(x)) = \gamma(x)$ for all x [6, 12]. α^0 is a function with maximal number of components such that $\alpha^0(f(x, u, d))$ does depend on d.

3   Basic Models and Problem Statement

Consider control system described by the equations

$$\begin{aligned} x(t+1) &= Fx(t) + Gu(t) + Dd(t) + \Psi(x(t), u(t)), \\ y(t) &= Hx(t), \end{aligned} \tag{4}$$

where

$$\Psi(x(t), u(t)) = C \begin{pmatrix} \varphi_1(A_1 x(t), u(t)) \\ \vdots \\ \varphi_q(A_q x(t), u(t)) \end{pmatrix},$$

F, G, H, C, and D are constant matrices, the functions $\varphi_1, \ldots, \varphi_q$ may be non-smooth, $A_1, \ldots, A_q$ are row matrices. The model (4) can be obtained from general nonlinear model (1) by some transformations described in [15].

When a fault occurs, $d(t)$ becomes an unknown function of time therefore the control problems for system (4) cannot be solved directly. To overcome this difficulty, we use a dynamic compensator similar to (2) described by

$$\begin{aligned} x_0^+ &= F_0 x_0 + G_0 u + J_0 y + C_0 \begin{pmatrix} \varphi_1(A_{01} z_0, u) \\ \vdots \\ \varphi_q(A_{0q} z_0, u) \end{pmatrix}, \\ u &= g(x_0, y, u_*), \end{aligned} \tag{5}$$

where the vector u_* is a new control, $x_0 \in R^k$, $k < n$, is the state vector of the compensator, $F_0, G_0, J_0, C_0, A_{01}, \ldots, A_{0q}$ are matrices to be determined, $z_0 =$

$\begin{pmatrix} x_0 \\ y \end{pmatrix}$. For simplicity, the notations x_0^+ and x_0 are used for $x_0(t+1)$ and $x_0(t)$, respectively. Denote the dynamic part of (5) by symbol S_0.

The goal of decoupling is to eliminate the effect of the unknown function $d(t)$ on a subsystem of maximal dimension of the closed-loop system (4), (5). Specifically, we are looking for system (5) and the subsystem S_* of the closed-looped system of the dimension $n_* < n$ as large as possible given by

$$x_*^+ = F_* x_* + G_* u_* + C_* \begin{pmatrix} \varphi_1(A_{*1}x_*, u_*) \\ \vdots \\ \varphi_q(A_{*q}x_*, u_*) \end{pmatrix}. \tag{6}$$

Clearly, subsystem (6) does not depend on the unknown function $d(t)$. As a result, one may find the control for subsystem (6) and solve the SRP. Note that since $n_* < n$, the fault effects can be eliminated only for some subvector of the state vector $x(t)$.

4 Disturbance Decoupling Problem

4.1 Main Relations

In this section, we assume that the function h_* is known and design the functions α and ξ. Then these results are used for solving the SRP. To simplify a solution, we assume that the functions h_*, α, and ξ are linear ones. This allows to solve the problem for nonlinear system (4) based on the methods of linear algebra. At first we consider the case when $q = 1$ and construct the system S_0 using so-called logic-dynamic approach [15].

In [5], α is found as a function with maximal number of components satisfying the condition $\alpha^0 \leq \alpha$. Since it is sought as a linear function, one assumes that

$$x_0 = \alpha(x) = \Phi x \tag{7}$$

for some matrix Φ of maximal rank satisfying the following conditions [1, 15]:

$$\begin{aligned} \Phi F &= F_0 \Phi + J_0 H, \\ G_0 &= \Phi G, \\ \Phi D &= 0. \end{aligned} \tag{8}$$

It can be shown that the additional relations $C_0 = \Phi C$ and

$$A = A_0 \begin{pmatrix} \Phi \\ H \end{pmatrix}, \tag{9}$$

corresponding to the nonlinear term, hold [13, 15]. Relation (9) is true if and only if rows of the matrix A linearly depend on the rows of the matrices Φ and H. Clearly, it is equivalent to the condition

$$rank \begin{pmatrix} \Phi \\ H \end{pmatrix} = rank \begin{pmatrix} \Phi \\ H \\ A \end{pmatrix}. \tag{10}$$

If $q > 1$, the matrix A in (9) and (10) is replaced by A_i, $i = 1, \ldots, q$.

It is assumed that the matrices F_0 and H_0 are sought in the form

$$F_0 = \begin{pmatrix} 0 & 1 & 0 & \cdots & 0 \\ 0 & 0 & 1 & \cdots & 0 \\ \vdots & \vdots & \vdots & \ddots & \vdots \\ 0 & 0 & 0 & \cdots & 0 \end{pmatrix},$$

$$H_0 = (1\ 0\ 0\ \cdots\ 0).$$

Then the equation $\Phi F = F_0 \Phi + J_0 H$ can be replaced by k equations:

$$\begin{aligned} \Phi_i F &= \Phi_{i+1} + J_{0i} H, \quad i = 1, \ldots, k-1, \\ \Phi_k F &= J_{0k} H, \end{aligned} \tag{11}$$

where Φ_i and J_{0i} are the ith rows of the matrices Φ and J_0, respectively, $i = 1, \ldots, k$, k is the dimension of the vector x_0.

4.2 Construction of the Dynamic Part of Compensator

Find the matrix Φ of maximal rank satisfying the condition $\Phi D = 0$. It was shown in [15] that (11) with the condition $\Phi D = 0$ can be transformed into the single equation

$$(\Phi_1 - J_{01}\ \cdots\ - J_{0k})\ (V^{(k)}\ B^{(k)}) = 0, \tag{12}$$

where

$$V^{(k)} = \begin{pmatrix} F^k \\ H F^{k-1} \\ \cdots \\ H \end{pmatrix}, \quad B^{(k)} = \begin{pmatrix} D & FD & \cdots & F^{k-1}D \\ 0 & HD & \cdots & HF^{k-2}D \\ \vdots & \vdots & \ddots & \vdots \\ 0 & 0 & \cdots & 0 \end{pmatrix}.$$

To construct the system S_0, set $k := n - p$ and check the condition

$$rank(V^{(k)}\ B^{(k)}) < lk + n. \tag{13}$$

When (13) holds, there exists the row $(\Phi_1 - J_{01} \cdots - J_{0k})$ such that (12) can be solved. Then one finds the matrix Φ from (11) and computes $G_0 := \Phi G$. Thus, the linear part of the system S_0 has been constructed.

If (13) does not hold, set $k := k - 1$ and repeat checking the condition (13). If (13) does not hold for all k, the system S_0 decoupled from $d(t)$ does not exist and the SRP has no solution. Because the dimension k is maximal, the function $\alpha(x) = \Phi x$ is the best choice for α in (3).

Note that if (10) is satisfied for the matrix Φ, the problem of the nonlinear system S_0 design reduces to the linear case: one finds the matrix A_0 from (9) and construct the dynamic part of the compensator (5). When (10) is not satisfied, find the set of N linearly independent solutions of (12) for some k and present them in the form

$$
\begin{aligned}
(\Phi_1^{(1)} - J_{01}^{(1)} \cdots - J_{0k}^{(1)}), \quad \cdots , \\
(\Phi_1^{(N)} - J_{01}^{(N)} \cdots - J_{0k}^{(N)}).
\end{aligned}
\tag{14}
$$

Theorem 2 ([13]) *Consider the matrices* $\Phi^{(1)}, \ldots, \Phi^{(N)}$ *obtained based on (11) and (12). Then the arbitrary linear combination of solutions from (14) with weight coefficients* $v_1, \ldots, v_N$ *gives the matrix*

$$
\Phi = v_1 \Phi^{(1)} + \cdots + v_N \Phi^{(N)}
$$

and describes some solution of our problem for the linear part of (5).

Assume that set of solutions of (12) is found in the form (14). To find the vector $v = (v_1, \ldots, v_N)$, represent (9) in the form

$$
A = A_{01} \Phi + A_{02} H,
\tag{15}
$$

where $A_0 = (A_{01} \ A_{02})$. Denote

$$
\Phi_1^{\Sigma} = \begin{pmatrix} \Phi_1^{(1)} \\ \vdots \\ \Phi_1^{(N)} \end{pmatrix}, \quad \ldots, \quad \Phi_k^{\Sigma} = \begin{pmatrix} \Phi_k^{(1)} \\ \vdots \\ \Phi_k^{(N)} \end{pmatrix},
$$

and rewrite (15) in the form

$$
A = A_{01} \begin{pmatrix} v\Phi_1^{\Sigma} \\ \vdots \\ v\Phi_k^{\Sigma} \end{pmatrix} + A_{02} H.
\tag{16}
$$

Analogously to (10), Eq. (16) has a solution if

$$rank \begin{pmatrix} \Phi^\Sigma \\ H \end{pmatrix} = rank \begin{pmatrix} \Phi^\Sigma \\ H \\ A \end{pmatrix}, \tag{17}$$

where

$$\Phi^\Sigma = \begin{pmatrix} \Phi_1^\Sigma \\ \vdots \\ \Phi_k^\Sigma \end{pmatrix}.$$

Assume that (17) holds and consider firstly the case when A is a row matrix. Here, (16) can be written in the form

$$A = (a_1 v \quad \ldots \quad a_k v)\Phi^\Sigma + A_{02} H,$$

where $(a_1 \quad \ldots \quad a_k) := A_{01}$, or in the form

$$A = A_v \Phi^\Sigma + A_{02} H, \tag{18}$$

where A_v is considered as unknown matrix. Solve (18) and find the matrices A_v and A_{02}. If A_v can be written in the form $(a_1 v \quad \ldots \quad a_k v)$ for some coefficients $a_1, \ldots, a_k$ and the vector $v = (v_1, \ldots, v_N)$, then stop, the matrices $A_{01} = (a_1 \quad \ldots \quad a_k)$ and A_{02} and the vector v of the weight coefficients have been obtained. Then the rows of matrices J_0 and Φ are found from the relations

$$J_{0j} = \sum_{i=1}^{N} v_i J_{0j}^{(i)}, \quad \Phi_j = \sum_{i=1}^{N} v_i \Phi_j^{(i)}, \quad j = 1, 2, \ldots, k;$$

$$G_0 = \Phi G, \quad C_0 = \Phi C.$$

As a result, a dynamic part of the compensator (5) has been built.

If the condition (17) is not true or the matrix A_v cannot be written in the form $(a_1 v, \ldots, a_k v)$, one has to decrease the dimension k and repeat the described procedure.

If the matrix A has several rows, Eq. (18) is solved for each rows with some coefficients $a_1, \ldots, a_k$ peculiar to the considered row; note that the vector v should be the same for all rows.

4.3 Construction of the Static Part of Compensator

Recall some results from [5]. Let $h_* = (h_{*1}, \ldots, h_{*L})^T$; denote by r_i and w_i the relative degrees of the function $h_{*i}(x)$ with respect to the input $u(t)$ and the unknown function $d(t)$, respectively. Besides, we use the notations

$$y_{*i}(k) = h_{*i}(x(k)) =: h_{*i,1}(x(k)), \dots,$$
$$y_{*i}(k + r_i - 1) =: h_{*i,r_i}(x(k)), \quad i = 1, \dots, L.$$

When $h_*(x) = H_* x$ for some matrix H_*, these relations reduce as follows.

Introduce the matrix C^*: if $C(i, j) \neq 0$ and the function φ_j contains components of the control vector u, set $C^*(i, j) = 1$, otherwise $C^*(i, j) = 0$.

Denote by r'_i minimal integer p such that $H_{*i} F^{p-1} G \neq 0$, by w_i minimal integer p such that $H_{*i} F^{p-1} D \neq 0$, and by r_i^* minimal integer p such that $H_{*i} F^{p-1} C^* \neq 0$, $i = 1, \dots, L$. Clearly, r'_i and r_i^* are the relative degrees of the output $y_{*i} = H_{*i} x$ with respect to $u(t)$ corresponding to the linear and nonlinear parts of system (4), respectively. Set $r_i := \min(r'_i, r_i^*)$, $i = 1, \dots, L$.

The following assumptions were made in [5].

Assumption 1 $w_i > r_i$ and $w_i > r'_i$ for all $i = 1, \dots, L$.

From the definition of r_i and Assumption 1,

$$y_{*i}(t + r_i) = \widehat{f}_i(x(t), u(t))$$

for some function $\widehat{f}_i$. Note that the function $\widehat{f}_i(x(t), u(t))$ is independent of the function $d(t)$ due to Assumption 1. Assuming $L \leq m$, we set

$$\widehat{f}(x, u) := (\widehat{f}_1(x, u), \dots, \widehat{f}_L(x, u))^T.$$

Vector $(r_1, \dots, r_L)$ is called a vector relative degree of y_* if

$$rank(\partial \widehat{f}(x, u)/\partial u) = L$$

everywhere except perhaps on a set of measure zero.

Assumption 2 The output y_* has a vector relative degree $(r_1, \dots, r_L)$.

Theorem 3 *[5] Under Assumptions 1 and 2, the controlled invariant function ξ with minimal number of components satisfying the inequality $\xi \leq h_*$, is computed by*

$$\xi := \begin{pmatrix} h_{*1}^0 \\ \vdots \\ h_{*L}^0 \end{pmatrix},$$

where $h_{*i}^0 = \left(h_{*i,1} \ \dots \ h_{*i,r_i} \right)^T$, $i = 1, \dots, L$.

To find the function ξ in a class of linear functions, we make the additional assumption.

Assumption 3 $r_i = r'_i$ for all $i = 1, \dots, L$, i.e. all relative degrees correspond to the linear part of system (4).

Set

$$y_{*1}^1 = H_{*1}x, \quad \ldots, \quad y_{*1}^{r_1} = H_{*1}F^{r_1-1}x + \psi_{r_1-1}^{(1)}(x);$$

clearly, the expression

$$y_{*1}^{r_1+} = H_{*1}F^{r_1-1}x^+ + \psi_{r_1-1}^{(1)}(x^+) = H_{*1}F^{r_1}x + H_{*1}F^{r_1-1}Gu + \psi_{r_1}^{(1)}(x)$$

is control dependent. Here $\psi_i^{(1)}(x)$, $i = 2, \ldots, r_1$, are some functions independent of the control due to Assumption 3. Clearly, the expression

$$H_{*1}F^{r_1}Fx + H_{*1}F^{r_1-1}Gu + \psi_{r_1}^{(1)}(x)$$

is similar to the function $\widehat{f}_1(x, u)$. By analogy, one obtains other expressions:

$$y_{*i}^{r_i+} = H_{*i}F^{r_i}Fx + H_{*i}F^{r_i-1}Gu + \psi_{r_i}^{(i)}(x), \quad i = 2, \ldots, L.$$

Consider the set of equations

$$\begin{cases} H_{*1}F^{r_1}x + H_{*1}F^{r_1-1}Gu + \psi_{r_1}^{(1)}(x) = u_{*1}, \\ \quad\quad \vdots \\ H_{*L}F^{r_L}x + H_{*L}F^{r_L-1}Gu + \psi_{r_L}^{(L)}(x) = u_{*L}. \end{cases} \tag{19}$$

Introduce the matrices

$$H_*^{(i)} = \begin{pmatrix} H_{*i} \\ \vdots \\ H_{*i}F^{r_i-1} \end{pmatrix}, \quad i = 1, \ldots, L, \quad \hat{H}_* = \begin{pmatrix} H_{*1}F^{r_1-1}G \\ \vdots \\ H_{*L}F^{r_L-1}G \end{pmatrix}. \tag{20}$$

If $r_i = \infty$, set $H_*^{(i)} := H_{*i}$. For simplicity, assume that $rank(\hat{H}_*) = L$ that is equivalent to Assumption 2. In this case (19) can be solved for the control u. Set

$$\Phi_* := \begin{pmatrix} H_*^{(1)} \\ \vdots \\ H_*^{(L)} \end{pmatrix}. \tag{21}$$

The matrix Φ_* is similar to the function ξ from Theorem 3, so Φ_* can be considered as controlled invariant for the linear part of the closed-loop system.

If the condition

$$rank\left(\Phi_*\right) = rank\begin{pmatrix} \Phi_* \\ A \end{pmatrix} \tag{22}$$

is true, then the nonlinear term in the system S_* can be constructed based on the linear part. Analogue of the condition $\alpha \leq \xi$ in (3) is the equality

$$rank\left(\varPhi\right) = rank\begin{pmatrix}\varPhi_* \\ \varPhi\end{pmatrix}. \tag{23}$$

If (22) and (23) are true, the SRP is solvable, otherwise a solution does not exist. Assume that (22) and (23) are true, then $\varPhi_* = Q\varPhi$ for some matrix Q.

Solving (19) for the control, one obtains the expression in the form $u = g'(x, u_*)$, i.e. a static state form of the feedback. Since $\varPhi_* = Q\varPhi$ and the matrix $\varPhi$ is similar to the (h, f)-invariant function, then the state x in $u = g'(x, u_*)$ can be expressed in the terms of the state $x_0 = \varPhi x$ and the vector y. As a result, a static part of the compensator (5) takes the form $u = g(x_0, y, u_*)$ for some function g.

The matrix $\varPhi_*$ is used to construct the system S_*: set $x_* = \varPhi_* x$ and compute

$$x_*^+ = \varPhi_* x^+ = \varPhi_* F x + \varPhi_* G u + \varPhi_* \varPsi(x, u). \tag{24}$$

Since $\varPhi_*$ is controlled invariant for the linear part of the closed-loop system and (23) is valid, the right-hand side in (24) can be expressed via the state $x_* = \varPhi_* x$ and the new input u_*, i.e. one obtains the expressions in the form (6).

If $r_i \neq r_i'$ for some i, then the function $\psi_{r_i}^{(i)}(x)$ in the left-hand sides in (19) depends on the control $u(t)$. In this case the matrix $\varPhi_*$ remains the analogue of ξ, but the expressions for g' and g become more complex [14].

5   Solution of the SRP

Here, we apply the solution of the DDP to solve the SRP. Note that in the statement of the SRP, the matrix H_* is a design object unlike the DDP where H_* is given. We are looking for a matrix H_* of maximal rank which does not depend on the function $d(t)$ using the compensator (5).

We take $H_* = \varPhi$ since this is the best choice for H_* due to Theorem 1 and check whether the DDP is solvable for such H_*. According to Theorem 1, a controlled invariant function ξ exist such that $\alpha \leq \xi \leq h_*$. Taking into account our analogues, one has $H_* = \varPhi$ and the only possible choice for $\varPhi_*$ is $\varPhi_* := \varPhi$. Hence, we have to check if $\varPhi$ is controlled invariant. If yes, the best solution of the SRP has been obtained and the system S_* has maximal possible dimension. Otherwise the SRP has a solution with the system S_* of smaller dimension. It is based on the matrix $\varPhi_*$ such that $\varPhi_* = Q\varPhi$. In this case $H_* := \varPhi_*$.

An algorithm computing the compensator (5) solving the SRP is given below.

Step 1. Construct the dynamic part of (5) based on the results of Sect. 4 and take $H_* := \varPhi$. Find the relative degrees r_i', r_i^*, and w_i of $y_{*i} = H_{*i} x$. Assuming that $r_i = r_i'$, check Assumption 1; if it does not hold, remove the ith row from the matrix $H_*, i = 1, \ldots, L$. Denote the final matrix by H_* as well.

Step 2. Compute the matrix $H_*^{(i)}$ from (20) and check the condition

Fig. 1 Control system

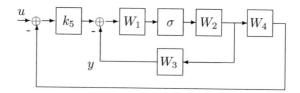

$$rank\left(\varPhi\right) = rank\left(\frac{H_*^{(i)}}{\varPhi}\right), \quad i = 1, \ldots, L. \qquad (25)$$

If it is not true, then remove the ith row from the matrix H_*. Denote the final matrix by H_* as well. Compute the matrix $\varPhi_*$ and check (21); if it is not true, then stop, the SRP is not solvable.

Step 3. Construct the matrix $\hat{H}_*$ from (20) and check whether $rank(\hat{H}_*) = L$. If yes, then go to Step 5, otherwise go to Step 4.

Step 4. Let $rank(\hat{H}_*) = M < L$. Find matrix P so that $rank(P\hat{H}_*) = M$. The matrix P collects the linearly independent rows in $\hat{H}_*$. Set $\hat{H}_* := P\hat{H}_*$.

Step 5. Construct the static part of (5) by solving (19) for u in the form $u = g'(x, u_*)$ and next in the form $u = g(x_0, y, u_*)$.

Step 6. Set $H_* := \varPhi_*$, $x_* := \varPhi_* x$, construct the system

$$x_*^+ = \varPhi_* x^+ = \varPhi_*(Fx + Gu + \Psi(x, u)),$$

and transform it into the form (6).

6 Example

Consider the control system shown in Fig. 1; it consists of linear subsystems with transfer functions

$$W_1(s) = k_1/(1 + T_1 s), \qquad W_2(s) = k_2/(1 + T_2 s),$$
$$W_3(s) = k_3 T_3 s/(1 + T_3 s), \quad W_4(s) = k_4/s,$$

and saturation operation

$$\sigma(z) = \begin{cases} z, & \text{if } |z| \le z_0, \\ z_0 \text{sign } z, & \text{if } |z| > z_0. \end{cases}$$

Note that this system was considered in [4] where the DDP was solved using methods of the functions' algebra. Now we solve the SRP by linear methods.

The Euler discretization of this system is described by the equations:

$$x_1^+ = kk_4x_2 + x_1,$$
$$x_2^+ = kk_2\sigma(x_3)/T_2 + x_2(1 - k/T_2),$$
$$x_3^+ = k(k_1k_5(u - x_1) - k_1k_3(x_2 - x_4))/T_1 + x_3(1 - k/T_1),$$
$$x_4^+ = kx_2/T_3 + x_4(1 - k/T_3) + kw,$$
$$y = k_3(x_2 - x_4).$$

Here $k_1 \div k_5$ are real coefficients, k is a discretization step, T_1 and T_2 are certain time constants; it is assumed that T_3 is subjected to the fault.

Matrices and nonlinearities describing the system are as follows:

$$F = \begin{pmatrix} 1 & kk_4 & 0 & 0 \\ 0 & 1 - k/T_2 & 0 & 0 \\ -kk_1k_5/T_1 & -k_1k_3 & 1 - k/T_1 & k_1k_3 \\ 0 & k/T_3 & 0 & 1 - k/T_3 \end{pmatrix}, \quad G = \begin{pmatrix} 0 \\ 0 \\ kk_1k_5/T_1 \\ 0 \end{pmatrix}, \quad (26)$$

$$H = k_3(0\ 1\ 0\ -1), \quad C = (0\ 1\ 0\ 0)^T, \quad D = (0\ 0\ 0\ 1)^T,$$
$$A = (0\ 0\ 1\ 0), \quad \varphi(x, u) = \sigma(Ax).$$

Compute the matrix D^0 of maximal rank such that $D^0 D = 0$:

$$D^0 = \begin{pmatrix} 1\ 0\ 0\ 0 \\ 0\ 1\ 0\ 0 \\ 0\ 0\ 1\ 0 \end{pmatrix}.$$

It can be shown that $\Phi = D^0$; clearly, (9) is satisfied therefore the nonlinear term of the compensator can be constructed based on the matrix Φ. Obviously, $C^* = 0$, $L = 3$. Next, by Step 1 of Algorithm find

$$r_1' = r_2' = \infty, \quad r_3' = 1, \quad r_1^* = r_2^* = r_3^* = \infty, \quad w_1 = w_2 = \infty, \quad w_3 = 2,$$

i.e. $w_3 > r_3 = min(r_3', r_3^*)$ and Assumptions 1 and 3 are satisfied.

Therefore $(x_{01}, x_{02}, x_{03})^T = (x_1, x_2, x_3)^T$ and description of the dynamic part of compensator is as follows:

$$x_{01}^+ = kk_4x_{02} + x_{01}$$
$$x_{02}^+ = kk_2\sigma(x_{03})/T_2 + x_{02}(1 - k/T_2) \quad (27)$$
$$x_{03}^+ = k(k_1k_5(u - x_{01}) - k_1y)/T_1 + x_{03}(1 - k/T_1).$$

Set $H_* := \Phi$. Obviously, $H_*^{(i)} = H_{*i}$ for all i and (25) is satisfied. Since $rank(\hat{H}_*) = 3$, Assumption 2 is true and one can set

$$u_* := k_1k_5(u - x_{01}) - k_1y.$$

As a result,

$$u = (u_* + k_1y)/(k_1k_5) + x_{01}.$$

Substituting u into (27), we obtain equations for the system S_*:

$$x_{*1}^+ = kk_4 x_{*2} + x_{*1}$$
$$x_{*2}^+ = kk_2 \sigma(x_{*3})/T_2 + x_{*2}(1 - k/T_2)$$
$$x_{*3}^+ = ku_*/T_1 + x_{*3}(1 - k/T_1),$$

where $(x_{*1}, x_{*2}, x_{*3})^T = (x_1, x_2, x_3)^T$.

Note that, in fact, the reconfiguration eliminates the internal loop in Fig. 1 and replaced it by a new control.

Acknowledgements This work was supported by grant of the Russian Scientific Foundation, projects 16-19-00046-P.

References

1. Alcorta-Garcia, E., Frank, P.: Deterministic nonlinear observer-based approach to fault diagnosis: a survey. Control Eng. Pract. **5**, 663–670 (1997)
2. Ding, S.: Data-Driven Design of Fault Diagnosis and Fault-Tolerant Control Systems. Springer-Verlag, London (2014)
3. Jiang, B., Staroswiecki, M., Cocquempot, V.: Fault accommodation for nonlinear dynamic systems. IEEE Trans. Autom. Control **51**, 1578–1583 (2006)
4. Kaldmäe, A., Kotta, Ü., Shumsky, A., Zhirabok, A.: Measurement feedback disturbance decoupling in discrete-time nonlinear systems. Automatica **49**, 2887–2891 (2013)
5. Kaldmäe, A., Kotta, Ü., Shumsky, A., Zhirabok, A.: Faulty plant reconfiguration based on disturbance decoupling methods. Asian J. Control **8**, 858–867 (2016)
6. Kaldmäe, A., Kotta, Ü., Shumsky, A., Zhirabok, A.: Disturbance decoupling in nonlinear hybrid systems. Nonlinear Anal. Hybrid Syst. **28**, 42–53 (2018)
7. Shumsky, A., Zhirabok, A., Jiang, B.: Fault accommodation in nonlinear and linear dynamic systems: fault decoupling based approach. Int. J. Innov. Comput. Inform. Control **7**, 4535–4549 (2011)
8. Staroswiecki, M., Yang, H., Jiang, B.: Progressive accommodation of aircraft actuator faults. In: IFAC Symposium Safe Process'2006, pp. 877–882. Beijing (2006)
9. Witczak, M.: Fault Diagnosis and Fault Tolerant Control Strategies for Nonlinear Systems. Springer, London (2014)
10. Yang, H., Jiang, B., Cocquempot, V.: Fault Tolerant Control Design for Hybrid Systems. Springer-Verlag, Berlin-Heidelberg (2010)
11. Zhang, K., Cocquempot, V.: Fast adaptive fault estimation and accommodation for nonlinear time-varing delay systems. Asian J. Control **11**, 643–652 (2009)
12. Zhirabok, A., Shumsky, A.: The Algebraic Methods for Analysis of Nonlinear Dynamic Systems (In Russian). Dalnauka, Vladivostok (2008)
13. Zhirabok, A., Shumsky, A.: Fault accommodation in nonlinear dynamic systems. J. Comput. Syst. Sci. Int. **48**, 559–566 (2009)
14. Zhirabok, A., Shumsky, A., Zuev, A., Bobko, E.: Faulty system reconfiguration: logic-dynamic approach. In: 12th IEEE Conference on Industrial Electronics and Applications, pp. 1479–1484. Siem Reap (2017)
15. Zhirabok, A., Shumsky, A., Solyanik, S., Suvorov, A.: Fault detection in nonlinear systems via linear methods. Int. J. Applied Math. Comput. Sci. **27**, 261–272 (2017)

Tri-valued Evaluation of Residuals as a Method of Addressing the Problem of Fault Compensation Effect

Jan Maciej Kościelny, Michał Bartyś, and Zofia Łabęda-Grudziak

Abstract The fault compensation effect sets an important and up-to-date unsolved theoretical and practical problem for diagnostics of processes. Compensation of influences of faults on residual values has a significant impact on the diagnosis. On the one hand, there is a view that model-based diagnostics does not allow for the fault isolation for which a fault compensation effect occurs. On the other hand, there is a view that application of approaches based on Reiter's theory allows for the isolation of such faults. The paper shows that these views are not necessarily true. Based on the analysis of a simple system composed of two interconnected tanks, it has been shown that the use of a tri-valued evaluation of residuals allows not only for increasing values of fault distinguishability metrics, but also to take into account the problem of their mutual compensation. In addition, it has been shown that consistency based diagnostics methods developed on the basis of Reiter's theory may lead to the generation of false diagnoses indicating seemingly compensating faults, even though it is, in fact, impossible due to specific physical limitations of the diagnosed system.

Keywords Fault isolation · Fault compensation effect · Consistency-based diagnostics · Fault distinguishability

J. M. Kościelny · M. Bartyś (✉) · Z. Łabęda-Grudziak
Institute of Automatic Control and Robotics, Warsaw University of Technology,
A. Boboli 8, 02-525 Warszawa, Poland
e-mail: michal.bartys@pw.edu.pl

J. M. Kościelny
e-mail: jan.koscielny@pw.edu.pl

Z. Łabęda-Grudziak
e-mail: labeda@mchtr.pw.edu.pl

© The Author(s), under exclusive license to Springer Nature Switzerland AG 2021
J. Korbicz et al. (eds.), *Advances in Diagnostics of Processes and Systems*,
Studies in Systems, Decision and Control 313,
https://doi.org/10.1007/978-3-030-58964-6_3

Fig. 1 Illustration for
exemplification of fault
compensation effect

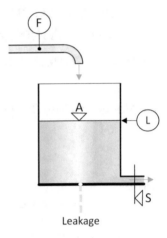

Leakage

1 Introduction

Multiple faults introduce fault masking effect. The fault masking takes place when
the presence of one fault may make it impossible to isolate some other faults or even
all of them. This takes place in model based diagnostics in cases where value or values
of residuals are invariant to the presence of multiple faults. In this particular case we
are talking about the effect of compensation of influences of faults on residuals. In the
course of the paper, this phenomenon will be called briefly as a fault compensation
effect.

In fact, the motivation for this work is that the fault compensation is still unsolved
theoretical and practical problem encountered in the diagnostics of processes. This
article proposes an approach to increase the fault distinguishability, despite fault
compensation effect by means of a tri-valued evaluation of residuals.

The problem of fault compensation will be narrowed down on a simple example.
Let us analyze the impacts of two chosen faults (pipe obliteration and tank leakage)
on the residuum value calculated from the flow balance equation derived for the
cylindric tank shown in Fig. 1. A liquid is feed into the tank by an inlet pipe. Assume
that inflow rate F of a liquid is measured. Similarly, the level L of the liquid in a
tank is measured by a level transducer. Therefore, residuum r is obtainable from:

$$r = F - \alpha \cdot S\sqrt{2gL} - A\frac{dL}{dt} , \tag{1}$$

where: α—outflow contraction coefficient, S—outflow pipe cross-section surface,
g—gravitational constant, A—cross-sectional area of the tank.

As can be inferred from Fig. 1, the two simultaneously occurring faults e.g. obliteration of the drain pipe (reduction of the cross-sectional area S) and a leak from the
tank have an adverse effect on the residuum value. Therefore, in a particular case,
they may have mutually compensating impact on the residual value r and therefore

residuum may takes value close to zero. In this case, the diagnostic signal determined on the basis of this residuum will be not indicative for any of both faults.

Note that fault compensation effect affects multiple faults. Therefore, the phenomenon of compensation can occur only if at least affects two faults to which particular residuum is sensitive. The close to zero value of residuum resulting from fault compensation adulterates value of fault symptom derived from this residuum. As a result, a set of observed symptoms may either match the signature of another diagnostic state of diagnosed system or does not match any signature at all. In the first case, the diagnosis will be false. In the second case, however, there will be a problem with the formulation of diagnosis. It should be mentioned that in case of threat, incorrect diagnose or lack thereof is unacceptable.

The most known methods of diagnostics were developed under assumption of single faults [1–4]. This assumption significantly simplifies fault isolation algorithms. However, it can be used for relatively non-complex systems, because over there, the probability of single faults is much greater than multiple ones. However, in the case of large scale complex industrial processes, such as, for example in: chemical, energy, food, etc. industries, the number of possible faults is relatively very high [5]. In these cases, the likelihood of multiple faults is also relatively high.

Effective consideration of the problem of fault compensation requires disposal of process models based on analytical relations between measured signals and faults. For the large-scale systems, acquiring of such models is too expensive and, in practice, usually not achievable. In addition, according to studies [6], diagnosis of multiple faults is multiply determined, depending on degree of practice, mental set, and the difficulty of the multiple fault itself.

In diagnostics of industrial processes, the only practical method of acquiring knowledge of the relations between faults and their symptoms is the use of expertise of technologists, process operators and automation engineers. The simplest form of synthesis of this knowledge is application of a binary diagnostic matrix (BDM) or set of conditional rules derived from this matrix [1–4]. The columns of the diagnostic matrix correspond to the signatures of faults, while the rows determine the subsets of faults to which individual binary diagnostic signals are sensitive.

To recognize multiple faults, we need to know the Boolean signatures of multiple faults. The signatures of system states can be derived directly from a binary diagnostic matrix. It is widely accepted [1, 5, 7, 8] that signature of diagnostic state is calculated as alternative of reference values of signatures of all single faults existing in this state. Note, however, that this method of determining signatures of multiple faults is not valid where the fault compensation effect may take place.

Compensation of mutual impacts of faults is referenced in consistency based methods based on Reiter's theory [9] and recognized as AI or DX methods [10–13]. These methods are also referred to as model-based diagnosis (MBD) [1, 14–17]. Diagnoses are generated as minimal hitting sets of all minimal conflict sets. The method has found so far applications for diagnosing of relatively simple systems [16, 17].

The conflict sets [9] refer to the faulty components of the system. Whilst these components are subject to be faulty, the conflict sets reflect somehow the sensitivity of results of diagnostic tests to the faults. Therefore, the conflict set corresponds to the set of faults denoted by values of logical 1s allocated in the binary diagnostic matrix in the row related with particular diagnostic test (observation). This is called as positive conflict. The generalized minimal hitting-set algorithm assuming more then two behavioral modes by component and nonpositive conflicts was proposed by Nyberg in [18].

Since the compensation of two faults occurs when their influence on the residuum is adverse, the natural way of taking into account and possibly addressing the problem of fault compensation, is the usage of signed tri-valued evaluation of residuals. This way of notion of relationship between faults and diagnostic signals is in fact qualitative, because it reflects the expert knowledge regarding the influence of fault on the residuum sign. It should be stressed that in case of availability of analytical models, the determination of this influence is quite trivial.

The tri-valued evaluation of residuals was considered in [19–21]. In these works, a Temporal Causal Graph (*TCG*) is used in order to create signatures of faults that reflect dynamics of the diagnosed system. However, this approach does not take multiple faults into consideration. Therefore, the effect of fault compensation can not be considered here at all.

Quite a lot of works were devoted to the application of the Fault Isolation System (*FIS*) [4, 22–25]. Generally, the *FIS* allows for inferring regarding faults based on relation between faults and multi-valued diagnostic signals. However, in these works the effect of residual compensation was also not taken into consideration.

Chiang et al. [26] proposed a three-step framework based on the modified distance and modified causal dependency to integrate the data-driven and causal connectivity based features with the propagation path-based feature for diagnosing known, unknown, and multiple faults. If faults contribute equally, the proposed algorithm classifies the multiple fault as unknown fault. When one fault is dominating over other in the masked multiple fault, the proposed algorithm diagnoses the dominating one.

Chiang's approach assumes that compensation of two faults leads to the isolation of unknown fault. This may be assumed as a drawback of this approach. We will show on example that tri-valued evaluation of residuals does not possesses this disadvantage.

According to the authors' best knowledge, the problem of compensation of faults for the case of tri-valued residuals was not analyzed so far.

2   Fault Isolation Based on Tri-valued Signatures

For a given diagnosed system, we can specify a set of possible faults:

$$F = \{f_k : k = 1, 2, \ldots, K\}. \tag{2}$$

Fault detection is carried out using a set of J tests. Each test generates a residuum r_j calculated either as the difference between the output of the model and the output of the diagnosed system or as a difference between both sides of the residuum equation:

$$R = \{r_j : j = 1, 2, \ldots, J\}. \tag{3}$$

Let us suppose that residuals will be the subject of a tri-valued assessment that distinguishes the signs of them. Therefore, each diagnostic signal may have three values $V_j = \{0, +1, -1\}$. The zero value of the diagnostic signal is interpreted as the insensitivity of the residuum to fault. This corresponds to the case where the residuum value is close to zero. The other two values of diagnostic signal are assumed as the symptoms of faults. The isolation of faults is carried out based on the knowledge of values of diagnostic signals from the set:

$$S = \{s_j : j = 1, 2, \ldots, J\}. \tag{4}$$

The relationship between diagnostic signal values and faults can be presented in the form of the Fault Isolation System (FIS) [4, 23, 24], which is essentially a generalization of the classic binary diagnostic matrix.

The FIS is an array structure consisting of K columns and J rows. Each structure element specifies a set of reference diagnostic signal values associated with each individual fault. Each column of the FIS structure defines a multi-valued signature $V(f_k)$ of each k-th fault, i.e. $V(f_k) = \left[V_{k1}, V_{k2}, \ldots, V_{kj} \ldots V_{kJ}\right]^T$, where V_{kj} is a subset of values of the diagnostic signal s_j corresponding to the fault f_k.

Please note that for some faults (e.g. instrument faults), residuals can react with a decrease or an increase in value, which means that both $\{-1\}$ and $\{+1\}$ values of a diagnostic signal are possible. Thus, in a tri-valued evaluation of residuals, the set of diagnostic signal values V_{kj} contains values belonging to subsets: $\{0\}$, $\{-1\}$, $\{+1\}$ and $\{-1, +1\}$.

The classic inference regarding faults is based on knowledge of values of all diagnostic signals. For this purpose, the actual diagnostic signal values are compared with the signatures of each diagnostic state i.e. full fitness, with single, double and eventually more multiple faults. However, there is a problem how we can specify signatures of states with multiple faults based on signatures of single faults only. In this paper, we will show an approach, addressed to double faults only which according to parsimony principle are most likely.

The rules for determining values of double faults signatures V_{jk} based on values of diagnostic signals of single faults V_j and V_k are shown in Table 1. These rules actually come down to a concise representation of the so-called alternative fault signatures. Generalization of reasoning with alternative signatures can be found in [27].

In the second column of Table 1, it is visible how the fault compensation effect influences the set of reference values of diagnostic signals in case of double faults. For each combination of reference values of single fault diagnostic signals that have

Table 1 The rules for determining values of double fault signatures

V_j / V_k	$-1/+1$	$-1/0$	$+1/0$	$-1/-1$	$+1/+1$	$0/0$
	$+1/-1$	$0/-1$	$0/+1$			
V_{jk}	$\{-1, 0, +1\}$	$\{-1\}$	$\{+1\}$	$\{-1\}$	$\{+1\}$	$\{0\}$

opposite signs, the set of double fault diagnostic signal reference values is addition-ally extended by the $\{0\}$ element. This not only increases the number of alternative signatures, but above all, gives the possibility to reflect the effect of compensation of faults in the process of diagnostic inference. This gives also the chance of increasing fault distinguishability. This aspect will be discussed further in this paper.

The diagnosis will be understood as indication of all states z_j of the system which signatures match the current diagnostic signal values.

$$DGN = \{z_i : V(z_i)\} = \left\{ z_j : \bigvee_{\forall j \in \{j=1,\dots,J\}} v_j \in V_j(z_i) \right\}, \tag{5}$$

where $V_j(z_i)$ is a subset of values of diagnostic signals s_j associated with signature of diagnostic state z_i.

3 Example

In further discussion, we will refer to a relatively simple example in the hope that this allow us to highlight the usefulness of a tri-valued residuals for diagnosing with respect to fault compensation effect. Let us consider diagnosing of a liquid fuel storage system. The idealized synoptic scheme of the system is shown in Fig. 2. Suppose that the fuel inflow rate F and fuel levels L_1 and L_2 in both tanks are available for diagnostic purposes. The list of potential faults are collected in Table 2. Let us further assume that the diagnosed system is critical due to the consequences of failures. Therefore, functional safe instrumentation (Safety Integrity Level SIL 3) is applied for the fuel tanks storage. In this case, we can assume that the probability of an instrument fault i.e. $\{f_5, f_6, f_7\}$ is negligible. Therefore, in further considerations their failures will not be taken into account. Obviously, in other circumstances, such faults must be unconditionally considered.

Applying Bernouli's law regarding incompressible and inviscid fluids we get:

$$A_1 \frac{dL_1}{dt} = F - \alpha_{12} \cdot S_{12}\sqrt{2g(L_1 - L_2)} \tag{6}$$

$$A_2 \frac{dL_2}{dt} = \alpha_{12} \cdot S_{12}\sqrt{2g(L_1 - L_2)} - \alpha_2 \cdot S_2\sqrt{2g \cdot L_2}, \tag{7}$$

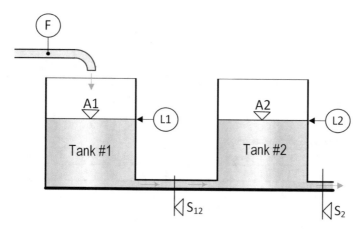

Fig. 2 The synoptic view of a fuel storage system. Notions: A_1 and A_2 are tank cross-sectional surfaces

Table 2 List of considered faults

Fault	Description
f_1	Obliteration of interconnection pipe
f_2	Obliteration of outlet pipe
f_3	Leaking tank 1
f_4	Leaking tank 2
f_5	Flow rate F measurement
f_6	Level L_1 measurement
f_7	Level L_2 measurement

where: S_{12}—cross-sectional area of the pipeline connecting tank 1 with tank 2; S_2—cross-sectional area of the outlet from tank 2; α_{12} and α_2—flow contraction coefficients.

For the sake of simplicity and clarity of the discussion, we will assume further that flow balance models (6) and (7) based on physical laws accurately reflect the dynamic properties of the diagnosed system in the state of fitness. On the basis of balance Eqs. (6) and (7), the following set of residuals is generated:

$$\begin{cases} r_1 = F - \alpha_{12} \cdot S_{12}\sqrt{2g(L_1 - L_2)} - A_1\frac{dL_1}{dt} \\ r_2 = \alpha_{12} \cdot S_{12}\sqrt{2g(L_1 - L_2)} - \alpha_2 \cdot S_2\sqrt{2g \cdot L_2} - A_2\frac{dL_2}{dt} \\ r_3 = F - \alpha_2 \cdot S_2\sqrt{2g \cdot L_2} - A_1\frac{dL_1}{dt} - A_2\frac{dL_2}{dt} \end{cases} \qquad (8)$$

A tri-valued sensitivity of given above residuals to faults is presented in Table 3 in the form of FIS information system [23, 24, 28].

Table 3 Fault isolation system for a two tanks system

F/S	f_1	f_2	f_3	f_4	f_5	f_6	f_7
s_1	+1	0	−1	0	−1, +1	−1, +1	−1, +1
s_2	−1	+1	0	−1	0	−1, +1	−1, +1
s_3	0	+1	−1	−1	−1, +1	−1+1	−1, +1

Table 4 Tri-valued signatures of diagnostic states

F/S	OK	f_1	f_2	f_3	f_4	$f_1 \wedge f_2$	$f_1 \wedge f_3$	$f_1 \wedge f_4$	$f_2 \wedge f_3$	$f_2 \wedge f_4$	$f_3 \wedge f_4$
s_1	0	+1	0	−1	0	+1	−1, 0 + 1	+1	−1	0	−1
s_2	0	−1	+1	0	−1	−1, 0 + 1	−1	−1	+1	−1, 0 + 1	−1
s_3	0	0	+1	−1	−1	+1	−1	−1	−1, 0 + 1	−1, 0 + 1	−1

On the basis of the knowledge of the signatures of single faults, we can now specify signatures of double faults according to rules specified in Table 1. Table 4 shows the signatures of the states of diagnosed system: fault free (OK), with single faults and with double faults. Here, the assumption regarding faultless instrumentation has been adopted.

The entries $\{-1, 0 + 1\}$ in Table 4 depict diagnostic states where the fault compensation effect can occur. From Table 4 follows the distinguisghability of majority of states. As it is easy to see, there are none unconditionally indistinguishable states [24, 29, 30]. Of the 11 diagnostic states concerned here, two states with single-faults and three states with double faults are distinguishable. In addition, together five states are conditionally distinguishable. For example, the combinations of diagnostic signal values given below lead to unambiguous diagnoses:

$$
\begin{aligned}
s_1 = -1, \; s_2 = 0, \; s_3 = -1 &\Rightarrow DGN = f_3 \\
s_1 = +1, \; s_2 = 0, \; s_3 = +1 &\Rightarrow DGN = f_1 \wedge f_2 \\
s_1 = +1, \; s_2 = -1, \; s_3 = 0 &\Rightarrow DGN = f_1 \\
s_1 = +1, \; s_2 = -1, \; s_3 = +1 &\Rightarrow DGN = f_1 \wedge f_2 \\
s_1 = +1, \; s_2 = +1, \; s_3 = +1 &\Rightarrow DGN = f_1 \wedge f_2
\end{aligned}
\tag{9}
$$

Conditionally distinguishable states occur for the combinations of diagnostic signals values collected in Table 5.

Let us discuss now some distinguishability related issues in the context of fault compensation effect.

For the double fault $\{f_2 \wedge f_4\}$, according to Table 4, the effect of compensation may occur in two tests s_2 and s_3. If so, then this state is not distinguishable from the OK state. It is worth mentioning that the probability of such event is very low. However, if it occurs, then diagnosis made within the proposed approach will be correct. None other approach does not provide this.

The state with fault f_4 is indistinguishable from the state with faults $\{f_1 \wedge f_3\}$ only in case of compensation of faults $\{f_1\}$ and $\{f_3\}$. Similarly, the lack of distin-

Table 5 Conditionally distinguishable diagnostic states

s_1	s_2	s_3	Conditionally distinguishable states
0	0	0	OK, $f_2 \wedge f_4$
0	-1	-1	f_4, $f_1 \wedge f_3$, $f_2 \wedge f_4$
0	$+1$	$+1$	f_2, $f_2 \wedge f_4$
-1	-1	-1	$f_3 \wedge f_4$, $f_1 \wedge f_3$
$+1$	-1	-1	$f_1 \wedge f_3$, $f_1 \wedge f_4$

guishability of state with fault $\{f_4\}$ from the state with faults $\{f_2 \wedge f_4\}$ occurs only when the effect of fault $\{f_4\}$ on the residual values r_2 and r_3 is dominant (diagnostic signals take $\{-1\}$ values) over the impact of fault $\{f_2\}$. If the fault $\{f_2\}$ for $(s_1 = 0,\ s_2 = s_3 = +1)$ is dominant, the diagnosis indicates states $\{f_2\}$ or $\{f_2 \wedge f_4\}$ states. In each of these cases, diagnoses are correct.

4 Comparison Study

4.1 Fault Isolation Based on a Binary Structure of Residual Sets

The binary signatures of each diagnostic state are specified commonly in a binary diagnostic matrix (BDM). The BDM matrix of fuel storage introduced in Sect. 3 is shown in Table 6. The reference values of diagnostic signals in states with double faults were determined under assumption that fault compensation effect does take place.

The only three states are distinguishable: OK, f_1, f_3. In contrast, the states $\{f_2, f_4, f_2 \wedge f_4\}$, $\{f_1 \wedge f_2, f_1 \wedge f_3, f_1 \wedge f_4, f_2 \wedge f_3, f_3 \wedge f_4\}$ are indistinguishable. As we can easily see, the fault distinguishability is reduced compared to FIS-based diagnosis discussed in previous section.

Diagnosing based on BDM as well as on FIS leads to indications of all states for which signatures match the actual values of diagnostic signals. Here, all values of diagnostic signals, i.e. fault symptoms and lacks of fault symptoms are taken into

Table 6 Bi-valued signatures of diagnostic states

F/S	OK	f_1	f_2	f_3	f_4	$f_1 \wedge f_2$	$f_1 \wedge f_3$	$f_1 \wedge f_4$	$f_2 \wedge f_3$	$f_2 \wedge f_4$	$f_3 \wedge f_4$
s_1	0	1	0	1	0	1	1	1	1	0	1
s_2	0	1	1	0	1	1	1	1	1	1	1
s_3	0	0	1	1	1	1	1	1	1	1	1

consideration. The inference is therefore carried out with assumption of exoneration of symptoms. The exoneration assumption means that a fault will always trigger all residuals sensitive to that fault [31].

4.2 Consistency Based Diagnostics

The concept of conflict sets is introduced in the consistency based diagnostics based on Reiter's theory [9, 14–17, 32]. Conflict set is meant as the set of components that contribute to malfunction of the diagnosed object. Therefore, a conflict set can be interpreted as the subset of faults to which a given diagnostic signal is sensitive. This approach takes into account only those conflicts that, in fact, have occurred. The absence of conflicts may result from the lack of faulty elements or due to delays of symptoms. Per analogy, in the diagnosing based on BMD or FIS it is possible to assume only fault symptoms and ignoring their absence.

Fault exoneration is frequently assumed in Fault Detection and Isolation (FDI) approaches based on analytic redundancy relations [11]. Under this assumption, if some relations are satisfied by observations, then all observation-related components are considered as non-faulty. In DX approaches, such assumption does not exist.

This means that the absence of conflict (zero value of diagnostic signal) cannot be the premise for reasoning of the health of elements from conflict set. In this case, the time sequence of symptoms is don't carry problem, obviously at the expense of lower fault distinguishability. Works [10, 32] indicate that Reiter's theory allows for classifying faults despite the effect of fault compensation. Table 7 provides a summary of potential diagnoses based on observed conflicts. The conflict sets in the analyzed example are as follows:

$$
\begin{aligned}
\{f_1, f_3\} & \quad corresponding\ with\ observation\ \{s_1 = 1\} \\
\{f_1, f_2, f_4\} & \quad corresponding\ with\ observation\ \{s_2 = 1\} \\
\{f_2, f_3, f_4\} & \quad corresponding\ with\ observation\ \{s_3 = 1\}.
\end{aligned}
\tag{10}
$$

The absence of observed conflicts indicates the state of fitness (OK). If there is only one conflict, the generated diagnosis indicates the entire conflict set (rows 2–4 in Table 7). It is therefore usually much less accurate than the diagnosis generated from the FIS information system, as well as from binary diagnostic matrix. It is worth mentioning that if a single symptom occurs, the faults in conflict sets are indistinguishable. If more than one conflict occurs, potential diagnoses indicate not only states with single, but also states with double faults (rows 5–8 in Table 7).

In row 5 of Table 7, potential diagnoses $\{f_2 \wedge f_3\}$, $\{f_3 \wedge f_4\}$ for $(s_1 = 1)$ and $s_2 = 1$ take into account the possibility of fault compensation by $(s_3 = 0)$. While the compensation may occur in the case of a potential diagnosis $\{f_2 \wedge f_3\}$, then in case of diagnosis $\{f_3 \wedge f_4\}$ it is not physically possible because both faults generate negative residuum values. This shows the weakness of Reiter's theory.

Table 7 Summary of potential diagnoses based on observed conflicts

Item	Conflict sets	s_1, s_2, s_3	Potential diagnoses
1	None	0, 0, 0	OK
2	$\{f_1, f_3\}$	1, 0, 0	$\{f_1\}, \{f_3\}$
3	$\{f_1, f_2, f_4\}$	0, 1, 0	$\{f_1\}, \{f_2\}, \{f_4\}$
4	$\{f_2, f_3, f_4\}$	0, 0, 1	$\{f_2\}, \{f_3\}, \{f_4\}$
5	$\{f_1, f_3\}, \{f_1, f_2, f_4\}$	1, 1, 0	$\{f_1\}, \{f_2 \wedge f_3\}, \{f_1 \wedge f_4\}$
6	$\{f_1, f_3\}, \{f_2, f_3, f_4\}$	1, 0, 1	$\{f_3\}, \{f_1 \wedge f_2\}, \{f_1 \wedge f_4\}$
7	$\{f_1, f_2, f_4\}, \{f_2, f_3, f_4\}$	0, 1, 1	$\{f_2\}, \{f_4\}, \{f_1 \wedge f_3\}$
8	$\{f_1, f_3\}, \{f_1, f_2, f_4\}, \{f_2, f_3, f_4\}$	1, 1, 1	$\{f_1 \wedge f_2\}, \{f_1 \wedge f_3\},$ $\{f_1 \wedge f_4\}, \{f_2 \wedge f_3\},$ $\{f_3 \wedge f_4\}$

Similarly, the potential diagnoses $\{f_1 \wedge f_2\}$, $\{f_1 \wedge f_4\}$ (row 6 of Table 7) for $(s_1 = 1)$ and $(s_3 = 1)$ assume compensation of these faults for $(s_2 = 0)$. It is not possible with faults $\{f_1 \wedge f_4\}$. On the other hand, for observations $(s_2 = 1)$ and $(s_3 = 1)$, the potential diagnosis $\{f_1 \wedge f_3\}$ presupposes compensation of these faults for $(s_1 = 0)$. Such compensation is physically possible.

This example indicates that only in some cases, the double faults may be assumed as correctly recognized based on methods derived from Reiter's theory. It takes place when faults cannot compensate their impacts on residuals because of physical limitations.

Remarks to Table 8:

1° Alternative diagnoses are formulated by a given set of diagnostic signal values.
2° The OK diagnoses in case of BDM and in consistency based approach do not take into account the possibility of compensation of faults $\{f_2 \wedge f_4\}$ in the case of absence of symptoms. This case is very unlikely.
3° BDM does not take into account the possibility of fault compensation. For example, for diagnostic signals: $(s_1 = 1)$, $(s_2 = 0)$, $(s_3 = 1)$ the compensation of faults $\{f_1 \wedge f_2\}$ in the residuum r_2 is not considered.
4° An advantage of the tri-valued residual evaluation over methods based on binary residual evaluation is clearly visible from Table 8 even without calculation of any fault distinguishability metric [28].

Table 8 shows a breakdown of diagnoses generated from binary diagnostic matrix, Reiter's theory, and FIS. Here, the signatures of diagnostic states with faults shown in Tables 4 and 6 are taken into account. However, in fact, other combinations of diagnostic signal values may occur due to false values of diagnostic signals caused by e.g. electromagnetic interference, measurement noise, inaccuracies of models used for fault detection, incorrect settings of thresholds, etc. This is a problem related to uncertainty of symptoms and uncertainty of inference with uncertainty. However, this problem goes beyond the framework of this paper.

Table 8 Collection of diagnoses based on BMD, FIS and Reiter's theory

Item	s_1, s_2, s_3	BDM	Consistency based	Tri-valued FIS	s_1, s_2, s_3
1	$0, 0, 0$	OK	OK	$OK, \{f_2 \wedge f_4\}$	$0, 0, 0$
2	$1, 1, 0$	$\{f_1\}$	$\{f_1\}, \{f_2 \wedge f_3\}$	$\{f_1\}$	$+1, -1, 0$
			$\{f_3 \wedge f_4\}$	$\{f_2 \wedge f_3\}$	$-1, +1, 0$
3	$1, 0, 1$	$\{f_3\}$	$\{f_3\}, \{f_1 \wedge f_2\}$	$\{f_1 \wedge f_2\}$	$+1, 0, +1$
			$\{f_1 \wedge f_4\}$	$\{f_3\}$	$-1, 0, -1$
4	$0, 0, 1$	$\{f_2\}$	$\{f_2\}$	$\{f_2\}, \{f_2 \wedge f_4\}$	$0, +1, +1$
		$\{f_4\}$	$\{f_4\}$	$\{f_2 \wedge f_4\}$	$0, +1, -1$
		$\{f_3 \wedge f_4\}$	$\{f_1 \wedge f_3\}$	$\{f_2 \wedge f_4\}$	$0, -1, +1$
				$\{f_4\}, \{f_1 \wedge f_3\}, \{f_2 \wedge f_4\}$	$0, -1, -1$
5	$1, 1, 1$	$\{f_1 \wedge f_2\}$	$\{f_1 \wedge f_2\}$	$\{f_1 \wedge f_2\}$	$+1, +1, +1$
		$\{f_1 \wedge f_3\}$	$\{f_1 \wedge f_3\}$	$\{f_1 \wedge f_2\}$	$+1, -1, +1$
		$\{f_1 \wedge f_4\}$	$\{f_1 \wedge f_4\}$	$\{f_1 \wedge f_2\}$	$+1, -1, -1$
		$\{f_2 \wedge f_3\}$	$\{f_2 \wedge f_3\}$	$\{f_2 \wedge f_3\}$	$-1, +1, +1$
		$\{f_3 \wedge f_4\}$	$\{f_3 \wedge f_4\}$	$\{f_2 \wedge f_3\}$	$-1, +1, -1$
				$\{f_2 \wedge f_3\}$	$-1, -1, +1$
				$\{f_1 \wedge f_3\}, \{f_3 \wedge f_4\}$	$+1, -1, -1$

5 Summary

The paper attempts to answer to some important questions related to fault compensation effect.

Firstly it was examined whether a three-valued evaluation of residuals could be useful for model based diagnosis, in cases where the compensation of mutual impacts of faults on residuals are likely. Based on example of diagnosing system consisting of two interconnected tanks it was shown that it is possible. Thus, this approach has features commonly attributed so far to family of DX based methods.

Secondly, the paper attempted to answer the question whether the possibility of distinguishing multiple faults assigned to consistency based methods is unconditional, i.e. ensures unconditional correctness of multiple fault isolation in any case where the effect of compensation of residuals takes place. The analysis performed showed that Reiter's theory based diagnostic methods can lead to false diagnoses indicating seemingly mutually compensating faults, even though their compensation is impossible due to physical limitations and the specificity of the diagnosed system.

Thirdly, it was shown on example that a tri-valued evaluation of residuals surely leads for achievement of better distinguishability of faults. This can be substantially important for the process safety. Precise diagnosis of critical faults is necessary for undertaking appropriate process protection actions. The lack of information of faults may result in incorrect decisions of process operators and in consequence introduces threats to the process. The findings obtained in this paper indicate the desirability of moving forwards the development of theory of fault isolation.

Acknowledgements The Authors acknowledge financing of this work by the grant No RPMA.01.02.00-IP.01-14-080/18 in the frames of European Fund of Regional Development.

References

1. Gertler, J.: Fault Detection and Diagnosis in Engineering Systems. Marcel Dekker Inc., New York, Basel, Hong Kong (1998)
2. Chen, J., Patton, R.: Robust Model Based Fault Diagnosis for Dynamic Systems. Kluwer Akademic Publishers, Boston (1999)
3. Patton, R., Frank, P., Clark, R. (eds.): Issues of Fault Diagnosis for Dynamic Systems. Springer-Verlag, Berlin, Heidelberg, New York (2000)
4. Korbicz, J., Kościelny, J.M., Kowalczuk, Z., Cholewa, W. (eds.): Fault Diagnosis. Models, Artificial Intelligence, Applications. Springer, Berlin (2004)
5. Kościelny, J.M., Bartyś, M., Syfert, M.: Methods of multiple fault isolation in large scale systems. IEEE Trans. Control Syst. Technol. **20**(5), 1302–1310 (2012)
6. Sanderson, P., Vernon, D., Reising, C., Augustiniak, M.: Effects of fault difficulty, expectancy, and prior exposure. Hum. Factors Ergon. Soc. **39**(9), 459–463 (1995). https://doi.org/10.1177/154193129503900904
7. Staroswiecki, M., Cassar, J.P., Declerck, P.: A structural framework for the design of FDI system in large scale industrial plants. In: Issues of Fault Diagnosis for Dynamic Systems, pp. 245–283. Springer (2000)
8. Bartyś, M.: Diagnosing multiple faults with dynamic binary matrix. IFAC-PapersOnLine **48**(21), 1297–1302 (2015). https://doi.org/10.1016/j.ifacol.2015.09.704
9. Reiter, R.A.: Theory of diagnosis from first principles. Artif. Intell. **32**(1), 57–95 (1987)
10. Cordier, M., Dague, P., Dumas, M., Lévy, F., Montmain, J., Staroswiecki, M., Travé-Massuyés, L.: Ai and automatic control approaches of model-based diagnosis: links and underlying hypotheses. In: 4th IFAC Symposium on Fault Detection Supervision and Safety for Technical Processes, Budapest, pp. 274–279 (2000)
11. Cordier, M., Dague, P., Lévy, F., Montmain, J., Staroswiecki, M., Travé-Massuyés, L.: Conflicts versus analytical redundancy relations: a comparative analysis of the model based diagnosis approach from the artificial intelligence and automatic control perspectives. IEEE Trans. Syst. Man Cybern. B: Cybern. **34**(5), 2163–2177 (2004)
12. Nyberg, M.: A fault isolation algorithm for the case of multiple faults and multiple fault types. In: 6th IFAC Safeprocess Symposium, Beijing, PR China, pp. 679–689 (2006)
13. Nyberg, M., Krysander, M.: Combining AI, FDI, and statistical hypothesis-testing in a framework for diagnosis. In: 5th IFAC Symposium on Fault Detection, Supervision and Safety of Technical Processes, Washington D.C., USA, June 9–11, pp. 891–896 (2003)
14. de Kleer, J., Williams, B.: Diagnosing multiple faults. Artif. Intell. **32**(1), 97–130 (1987)
15. de Kleer, J., Kurien, J.: Fundamentals of model-based diagnosis. In: 5th IFAC Symposium on Fault Detection, Supervision and Safety of Technical Processes, Washington, D.C., USA, June 9–11, pp. 25–36 (2003)
16. Ng, H.T.: Model-based, multiple-fault diagnosis of dynamic, continuous physical devices. IEEE Expert **6**, 38–43 (1991)
17. Górny, B.: Consistency-based reasoning inmodel-based diagnosis. Ph.D. thesis, Akademia Górniczo-Hutniczaw Krakowie, Wydział Elektrotechniki, Automatyki, Informatyki iElektroniki (2002)
18. Nyberg, M.: A generalized minimal hitting-set algorithm to handle diagnosis with behavioral modes. IEEE Trans. Syst. Man Cybern. Part A: Syst. Hum. **41**(1), 137–148 (2011). https://doi.org/10.1109/TSMCA.2010.2048750
19. Biswas, G., Kapadia, R., Yu, X.: Combined qualitative-quantitative steady-state diagnosis of continuous-valued systems. IEEE Trans. Syst. Man Cybern. Part A: Syst. Hum. **27**(2), 167–185 (1997)

20. Daigle, M., Koutsoukos, X., Biswas, G.: A qualitative approach to multiple fault isolation in continuous systems. In: Proceedings of the National Conference of Artificial Intelligence, vol. 22 (2007)
21. Daigle, M., Koutsoukos, X., Biswas, G.: A qualitative event-based approach to continuous systems diagnosis. IEEE Trans. Control Syst. Technol. **17**(4), 780–793 (2009)
22. Kościelny, J.M.: Application of fuzzy logic fault isolation in a three-tank system. In: IFAC Proceedings Volumens, vol. 32, 14 World Congress IFAC, Bejing, China, pp. 7754–7759 (1999)
23. Kościelny, J.M.: Diagnostyka zautomatyzowanych procesów przemysłowych (in polish). Akademicka Oficyna Wydawnicza Exit, Warszawa (2001)
24. Kościelny, J.M., Bartyś, M., Rzepiejewski, P., da Costa, J.M.G.S.: Actuator fault distinguishability study of the damadics benchmark problem. Control Eng. Pract. **14**(6), 645–652 (2006)
25. Kościelny, J.M., Syfert, M.: Fuzzy diagnostic reasoning that takes into account the uncertainty of the faults-symptoms relation. Int. J. Appl. Math. Comput. Sci. **16**(1), 27–35 (2006)
26. Chiang, L., Jiang, B., Zhu, X., Huang, D., Braatz, R.: Diagnosis of multiple and unknown faults using the causal map and multivariate statistics. J. Process Control **28**, 27–39 (2015)
27. Bartyś, M.: Generalised reasoning about faults based on diagnostic matrix. Int. J. Appl. Math. Comput. Sci. **23**(2), 407–417 (2013)
28. Kościelny, J.M., Bartyś, M., Rostek, K.: The comparison of fault distinguishability approaches case study. Bull. Pol. Acad. Sci. Tech. Sci. **67**(6), 1059–1068 (2019). https://doi.org/10.24425/bpasts.2019.131566
29. Kościelny, J.M., Syfert, M., Rostek, K., Sztyber, A.: Fault isolability with different forms of faults-symptoms relation. Int. J. Appl. Math. Comput. Sci. **26**(4), 815–826 (2016)
30. Kościelny, J.M., Syfert, M., Sztyber, A.: Rozróżnialność uszkodzeń w diagnostyce złożonych obiektów dynamicznych. EXIT, Warszawa (2019)
31. Jung, D., Khorasgani, H., Frisk, E., Krysander, M., Biswas, G.: Analysis of fault isolation assumptions when comparing model-based design approaches of diagnosis systems. In: Preprints of 9th IFAC Symposium on Fault Detection, Supervision and Safety of Technical Processes, Paris, France, September 2–4, pp. 1289–1296 (2015)
32. Travé-Massuyés, L.: Bridges between diagnosis theories from control and AI perspectives. In: Intelligent Systems in Technical and Medical Diagnostics, vol. 230, pp. 441–452. Springer, Heidelberg (2014)

Leader-Following Formation Control for Networked Multi-agent Systems Under Communication Faults/Failures

Juan Antonio Vazquez Trejo, Didier Theilliol, Manuel Adam Medina, C. D. García Beltrán, and Marcin Witczak

Abstract In this paper, a leader-following formation control for second-order multi-agent systems under communication faults/failures is presented. Communication faults/failures are considered dependent on the distance between the agents as smooth time-varying delays. Linear matrix inequality conditions are obtained for computing a control gain which guarantee stability when communication faults occur. Under the framework of control synthesis applied to time-varying delay multi-agent systems, the main contribution in this paper is the design of a leader-following formation control for second-order multi-agent systems such that all the faulty agents achieve the trajectories of a virtual agent and maintain a desired formation despite communication faults. A comparison between the classical formation control and our strategy is considered through numerical examples.

Keywords Multi-agent systems · Consensus · Formation control · Fault/failure communications

J. A. Vazquez Trejo · D. Theilliol (✉)
University of Lorraine, CRAN UMR 7039 CNRS, Nancy, France
e-mail: didier.theilliol@univ-lorraine.fr

J. A. Vazquez Trejo
e-mail: juan-antonio.vazquez-trejo@univ-lorraine.fr; vtjuan@cenidet.tecnm.mx

J. A. Vazquez Trejo · M. Adam Medina · C. D. García Beltrán
TECNM/CENIDET Electronic Engineering Department, Cuernavaca, Morelos, Mexico
e-mail: manuel.am@cenidet.tecnm.mx

M. Witczak
Institute of Control and Computation Engineering, University of Zielona Góra, ul. Szafrana 2, 65-516 Zielona Góra, Poland
e-mail: m.witczak@issi.uz.zgora.pl

© The Author(s), under exclusive license to Springer Nature Switzerland AG 2021
J. Korbicz et al. (eds.), *Advances in Diagnostics of Processes and Systems*,
Studies in Systems, Decision and Control 313,
https://doi.org/10.1007/978-3-030-58964-6_4

1 Introduction

Multi-agent systems have been of interest due to their potential accomplishing missions that a single agent cannot perform. The main characteristic of multi-agent systems is the exchange of information through digital networks. Nevertheless, bandwidth limitation, delays, or packet losses are present challenges in real engineering applications for multi-agent systems [1]. A stochastic leader-following consensus control of single integrator multi-agent systems with additive, multiplicative noises and time-delay is addressed in [2]. A mean square leader-following consensus for multi-agent systems with measurement noises and time-varying delays is investigated in [3]. A leaderless and leader-following consensus for a class of multi-agent chaotic systems with unknown time delays under switching topologies is investigated in [4]. In [5], a leader-following consensus for multiple uncertain Euler-Lagrange systems subject communication delays and switching networks is presented. In [6], a self-triggered algorithm for leader-following consensus with input time-delay is proposed. In [7], an event-triggered leader-following consensus for general linear fractional-order multi-agent systems subject to input delay is investigated. A mean square leader-following consensus for stochastic multi-agent systems using a distributed event-triggered mechanism with time-varying delay is addressed in [8]. An adaptive leader-following consensus using a centralized event-triggered function for a class of heterogeneous first-order multi-agent systems is studied in [9]. A reduced-order observer-based consensus for multi-agent systems with time delay an event-triggered strategy is investigated in [10]. An event-triggered containment control for first-order and second-order multi-agent systems with constant delays is presented in [11].

In the aforementioned works, robust strategies for different types of multi-agent systems are presented with constant delays, time-varying delays, stochastic or deterministic, with switching networks, noises or disturbance. In this paper, a control gain is designed in order to guarantee stability and to follow a virtual agent when communication faults occur in one or all agents extending [12] to leader-following multi-agent systems. Unlike [12] where a robust state feedback controller for the generalized time-delay systems with delayed states is considered, the main contribution in this paper is the design of a leader-following formation control for second-order multi-agent systems under communication faults such that all agents achieve the trajectories of a virtual agent and maintain a desired formation.

The paper is organized as follows. In Sect. 2, notation, graph theory, and the problem statement are contained. In Sect. 3, the leader-following formation control design under communication faults/failures is presented. In Sect. 4, simulation results are presented to illustrate the strategy effectiveness. Finally, in Sect. 5, conclusions are presented.

2 Preliminaries and Problem Statement

2.1 Graph Theory and Notation

A matrix X, X^T denotes its transpose, $X > 0 (< 0)$ denotes a symmetric positive (negative) definite matrix. The set $\mathbb{R}$ denotes the real numbers. $\|.\|$ denotes the Euclidean norm. For simplicity, the symbol $*$ within a symmetric matrix represents the symmetric entries. The Hermitian part of a square matrix X is denoted by $\text{He}\{X\} := X + X^T$. The column vector $\mathbf{1}_N$ is a column vector of N unity elements. The symbol $\otimes$ denotes the Kronecker product.

Lemma 1 ([13]) *For real matrices $\bar{A}$, $\bar{B}$, $\bar{C}$, and $\bar{D}$ with appropriate dimensions, the Kronecker product $\otimes$ has the following properties:*

(1) $(\bar{A} + \bar{B}) \otimes \bar{C} = \bar{A} \otimes \bar{C} + \bar{B} \otimes \bar{C}$,
(2) $(\bar{A} \otimes \bar{B})^T = \bar{A}^T \otimes \bar{B}^T$,
(3) $(\bar{A} \otimes \bar{B})(\bar{C} \otimes \bar{D}) = (\bar{A}\bar{C}) \otimes (\bar{B}\bar{D})$.

Lemma 2 ([14]) *For a given symmetric matrix $\begin{bmatrix} \bar{A} & \bar{B} \\ \bar{B}^T & \bar{C} \end{bmatrix} < 0$ the following statements are equivalent:*

(1) $\bar{A} < 0$, $\bar{C} - \bar{B}^T \bar{A}^{-1} \bar{B} < 0$,
(2) $\bar{C} < 0$, $\bar{A} - \bar{B}^T \bar{C}^{-1} \bar{B} < 0$.

A directed graph $\mathcal{G}$ is a pair $(\mathcal{V}, \mathcal{E})$, where $\mathcal{V} = \{\mathbf{v}_1, \ldots, \mathbf{v}_N\}$ is a non-empty finite node set and $\mathcal{E} = \{(i, j) : i, j \in \mathcal{V}\} \subseteq \mathcal{V} \times \mathcal{V}$ is an edge set of ordered pairs of nodes, called edges. N is the number of all nodes. The neighbors of node i are denoted as $j \in \mathcal{N}_i$. The adjacency matrix $\mathcal{A} = [a_{ij}] \in \mathbb{R}^{N \times N}$ associated with the graph $\mathcal{G}$ is defined such that $a_{ii} = 0$, $a_{ij} > 0$ if and only if $(i, j) \in \mathcal{E}$ and $a_{ij} = 0$ otherwise. The adjacency matrix of an undirected graph (bidirectional communication) is defined as $a_{ij} = a_{ji}$ for all $i \neq j$ and $\mathcal{A} = \mathcal{A}^T$. The Laplacian matrix $\mathcal{L} = [\mathcal{L}_{ij}] \in \mathbb{R}^{N \times N}$ of the graph $\mathcal{G}$ is defined as $\mathcal{L}_{ii} = \sum_{j \neq i} a_{ij}$ and $\mathcal{L}_{ij} = -a_{ij}$, $i \neq j$.

2.2 Problem Statement

Consider a double integrator multi-agent system of N agents:

$$\begin{aligned} \dot{p}_i(t) &= v_i(t), \\ \dot{v}_i(t) &= u_i(t), \end{aligned} \tag{1}$$

$p_i(t)$, $v_i(t)$, $u_i(t) \in \mathbb{R}^n$ are respectively the position, velocity, and acceleration input ($\forall i = 1, 2, \ldots, N$) in a n dimensional Euclidean space.

Assumption 1 The graph $\mathcal{G}$ is undirected and connected.

Lemma 3 ([15]) *The Laplacian matrix $\mathcal{L}$ associated with an undirected graph $\mathcal{G}$ has at least one zero eigenvalue and all of the nonzero eigenvalues are real positive, $0 = \lambda_1(\mathcal{L}) \leq \lambda_2(\mathcal{L}) \leq \cdots \leq \lambda_N(\mathcal{L})$. The Laplacian matrix $\mathcal{L}$ has exactly one zero eigenvalue if and only if $\mathcal{G}$ is connected and the eigenvector associated with the eigenvalue zero is 1_N.*

Let us define a virtual leader described by $p_r(t)$ and $v_r(t)$, where $p_r(t)$ and $v_r(t) \in \mathbb{R}^n$ are respectively the position and velocity of a virtual agent. From the agent i and its neighbors j, let us define h_i and $h_j \in \mathbb{R}^n$ as the rigid desired-position formation. The classical leader-following formation control reported in [16, 17] under communication faults/failures is defined as follows:

$$
\begin{aligned}
u_i(t) = &\sum_{j \in \mathcal{N}_i} a_{ij} \left[p_j(t - \tau_{ij}(t)) - p_i(t - \tau_{ij}(t)) - (h_j - h_i) \right] \\
&+ \sum_{j \in \mathcal{N}_i} a_{ij} \left[v_j(t - \tau_{ij}(t)) - v_i(t - \tau_{ij}(t)) \right] - (v_i(t) - v_r(t))
\end{aligned}
\tag{2}
$$

where $\mathcal{N}_i$ is the set of i's neighbors, $\tau_{ij}(t)$ is a communication fault/failure between agent i and its neighbors [2–5].

Assumption 2 All the agents have information of the virtual leader states.

Definition 1 The agents in (1) under (2) reach the formation in a consensus, if:

$$
\lim_{t \to \infty} \left\| (p_i(t) - h_i(t)) - (p_j(t) - h_j(t)) \right\| \to 0,
\tag{3}
$$

$$\forall i = 1, \ldots, N.$$

Remark 1 When $\tau_{ij}(t) = 0$, the leader-following formation control can be solved as reported in [16] if and only if $\mathcal{G}$ contains a spanning tree. When $\tau_{ij}(t) > 0$, the longest delay to reach a consensus is determined in [15] as $\tau_{ij} < \pi / (2\lambda_N(\mathcal{L}))$, where $\lambda_N(\mathcal{L})$ is the maximum eigenvalue of the Laplacian matrix. In [15], the delay is considered constant and with the same value for all agents ($\tau_{ij}(t) = \tau$).

In this paper, communication faults are considered dependent on the agent positions. The communication faults are described by the following function:

$$
\tau_{ij}(t) = \left(\gamma - \gamma e^{-\beta_1 \| p_i(t) - p_j(t) \|} \right) \left(0.5 + 0.5 \tanh \left(\beta_2 (t - t_f) \right) \right),
\tag{4}
$$

where γ, β_1, and β_2 are positive scalars; t_f is the time at which the fault occurs. Note that, $\tau_{ij}(t) = \gamma$ is the maximum value approximately. Communication faults can be associated to a degradation of the communication between agents in link with their distance. Similarly, such assumption has been considered in stochastic approach [18].

Assumption 3 $\dot{\tau}_{ij}(t) \leq d_1 < 1, \forall i \neq j, j \in \mathcal{N}_i$.

The focus of this paper is to design a leader-following formation control under communication faults for the second-order multi-agent system (1) such that all the agents follow the trajectories of a virtual agent and maintain a desired formation. In the following Section, the design of a leader-following formation control under communication faults is presented.

3 Leader-Following Formation Control Design Under Communication Faults: Proposed Approach

Let us define the tracking error between the agent i and the virtual agent $\bar{p}_i(t) = p_i(t) - p_r(t)$, $\bar{v}_i(t) = v_i(t) - v_r(t)$. Let $\delta_i(t) = \left[\bar{p}_i(t)^T - h_i^T, \bar{v}_i(t)^T\right]^T$. The tracking error dynamic is defined as follows:

$$\dot{\delta}_i(t) = A\delta_i(t) + Bu_i(t), \text{ with } A = \begin{bmatrix} 0_{n\times n} & I_n \\ 0_{n\times n} & 0_{n\times n} \end{bmatrix}, B = \begin{bmatrix} 0_{n\times n} \\ I_n \end{bmatrix}. \tag{5}$$

The control (2) is modified as follows:

$$u_i(t) = \sum_{j\in\mathcal{N}_i} a_{ij} \left[\bar{p}_j(t - \tau_{ij}(t)) - \bar{p}_i(t - \tau_{ij}(t)) - (h_j - h_i)\right]$$
$$+ \sum_{j\in\mathcal{N}_i} a_{ij} \left[\bar{v}_j(t - \tau_{ij}(t)) - \bar{v}_i(t - \tau_{ij}(t))\right] - \bar{v}_i(t) \tag{6}$$

Based on change coordinates (5) and adding the control gain K_c and the scalar α, the proposed leader-following formation control under communication faults/failures when $\tau_{ij}(t) > 0$ is defined as follows:

$$u_i(t) = K_c \left[\sum_{j\in\mathcal{N}_i} a_{ij} \left(\delta_i(t - \tau_{ij}(t)) - \delta_j(t - \tau_{ij}(t))\right) + \alpha\delta_i(t)\right] \tag{7}$$

where $K_c \in \mathbb{R}^{n\times 2n}$ is the control gain to be designed and $\alpha > 0$ must be a positive constant. Note that, if $K_c = \begin{bmatrix} -I_n & -I_n \end{bmatrix}$ and $\alpha = 1$, the classical formation control (2), it recovers. Based on (7), (5) is equivalent to:

$$\dot{\delta}_i(t) = A\delta_i(t) + BK_c \left[\sum_{j\in\mathcal{N}_i} a_{ij} \left(\delta_i(t - \tau_{ij}(t)) - \delta_j(t - \tau_{ij}(t))\right) + \alpha\delta_i(t)\right]. \tag{8}$$

Let $\delta(t) = \left[\delta_1(t)^T, \delta_2(t)^T, \dots, \delta_N(t)^T\right]^T$ and $\delta(t - \tau(t)) = \left[\delta_1(t - \tau_{1j}(t))^T, \delta_2(t - \tau_{2j}(t))^T, \dots, \delta_N(t - \tau_{Nj}(t))^T\right]^T$, then, the tracking error model (8) is rewritten as follows:

$$\dot{\delta}(t) = (I_N \otimes (A + \alpha B K_c)) \delta(t) + (\mathcal{L} \otimes B K_c) \delta(t - \tau(t)). \tag{9}$$

The following theorem provides LMI conditions for the computation of the control gain K_c.

Theorem 1 *Considering the closed-loop system in (9), given the non-zero eigenvalues λ_i ($\mathcal{L}$) of the Laplacian matrix ($i = 2, 3, \dots, N$), the scalar $\alpha > 0$, scalars $\mu_1 > 0$, $\mu_2 > 0$, and $\dot{\tau}_{ij} \leq d_1 < 1$; if there exist matrices $P_1 = P_1^T > 0$, $P_2 = P_2^T > 0$, and K_c such that the inequality (10) holds, then the leader-following formation control problem for the system (1) under communication faults/failures (4) is quadratically stable under (7).*

$$\begin{bmatrix} \mathrm{He}\,\{P_1 A\} + I - \frac{2P_1}{\mu_1} + P_2 & 0 & -\lambda_i P_1 B \frac{P_1}{\mu_1} + \mu_1 \alpha B K_c \\ * & -(1 - d_1) P_2 & -K_c^T & 0 \\ * & * & -2\mu_2 I & 0 \\ * & * & * & -I \end{bmatrix} < 0 \tag{10}$$

Proof Let us define a Lyapunov functional as follows:

$$V = \delta(t)^T (I_N \otimes P_1) \delta(t) + \int_{t-\tau}^{t} \delta(s)^T (I_N \otimes P_2) \delta(s) ds \tag{11}$$

The derivative V along the trajectories of (9) is given by:

$$\dot{V} = 2\delta(t)^T (I_N \otimes P_1) \dot{\delta}(t) + \delta(t)^T (I_N \otimes P_2) \delta(t) \\ - (1 - \dot{\tau}(t))\delta(t - \tau(t))^T (I_N \otimes P_2) \delta(t - \tau(t)). \tag{12}$$

According to [12], (12) is negative-definite when

$$\dot{V} = 2\delta(t)^T (I_N \otimes P_1) \dot{\delta}(t) + \delta(t)^T (I_N \otimes P_2) \delta(t) \\ - (1 - d_1)\delta(t - \tau(t))^T (I_N \otimes P_2) \delta(t - \tau(t)) < 0. \tag{13}$$

where $\dot{\tau}(t) \leq d_1 < 1$ according to Assumption 3. Thus,

$$\dot{V} = 2\delta(t)^T (I_N \otimes (P_1 A + \alpha P_1 B K_c)) \delta(t) + 2\delta(t)^T (\mathcal{L} \otimes P_1 B K_c) \delta(t - \tau(t)) \\ + \delta(t)^T (I_N \otimes P_2) \delta(t) - (1 - d_1)\delta(t - \tau(t))^T (I_N \otimes P_2) \delta(t - \tau(t)). \tag{14}$$

Let us perform a spectral decomposition, such that $\mathcal{L} = TJT^T$ with an orthogonal matrix $T \in \mathbb{R}^{N \times N}$ and a diagonal matrix $J = diag\,(0, \lambda_1, \lambda_2, \ldots, \lambda_N) \in \mathbb{R}^{N \times N}$. Define a change of coordinates as follows:

$$\varphi(t) = \left(T^T \otimes I_N\right) \delta(t),$$
$$\varphi(t - \tau(t)) = \left(T^T \otimes I_N\right) \delta(t - \tau(t)). \tag{15}$$

Replacing (15) in (14) is rewritten as follows:

$$\dot{V} = 2\varphi(t)^T \left(I_N \otimes (P_1 A + \alpha P_1 B K_c)\right) \varphi(t) + 2\varphi(t)^T \left(J \otimes P_1 B K_c\right) \varphi(t - \tau(t))$$
$$+ \varphi(t)^T \left(I_N \otimes P_2\right) \varphi(t) - (1 - d_1)\varphi(t - \tau(t))^T \left(I_N \otimes P_2\right) \varphi(t - \tau(t)). \tag{16}$$

By Lemma 3, it is obtained $\varphi_1(t) = 0$ and $\varphi_1(t - \tau(t)) = 0$ due to $\lambda_1 = 0$, then (16) can be rewritten as follows:

$$\dot{V} = \sum_{i=2}^{N} \varphi_i(t)^T \mathrm{He}\,\{P_1 A + \alpha P_1 B K_c\} \varphi_i(t) + 2 \sum_{i=2}^{N} \varphi_i(t)^T \left(\lambda_i P_1 B K_c\right) \varphi_i(t - \tau)$$
$$+ \sum_{i=2}^{N} \varphi_i(t)^T \left(P_2\right) \varphi_i(t) - (1 - d_1) \sum_{i=2}^{N} \varphi_i(t - \tau)^T \left(P_2\right) \varphi_i(t - \tau). \tag{17}$$

Then, it is obtained:

$$\dot{V} = \sum_{i=2}^{N} \left[\varphi_i(t)\ \varphi_i(t - \tau)\right] \Theta_i \begin{bmatrix} \varphi_i(t) \\ \varphi_i(t - \tau) \end{bmatrix} < 0,$$
$$\Theta_i = \begin{bmatrix} \mathrm{He}\,\{P_1 A + \alpha P_1 B K_c\} + P_2 & \lambda_i P_1 B K_c \\ * & -(1 - d_1) P_2 \end{bmatrix}. \tag{18}$$

If $\Theta_i < 0$ is definite-negative $\forall i = 2, 3, \ldots, N$, then, $\dot{V} < 0$. Thus, the leader-following formation control problem for the system (1) under communication faults/failures is quadratically stable under (7). The inequality (18) guarantees the asymptotic stability, however, the matrices cannot be calculated using conventional tools. Using Schur complement in (10), it is obtained:

$$\begin{bmatrix} Q_1 & 0 & -\lambda_i P_1 B \\ * & -(1 - d_1) P_2 & -K_c^T \\ * & * & -2\mu_2 I \end{bmatrix} < 0. \tag{19}$$

where $Q_1 = \mathrm{He}\,\{P_1 A\} + I - \frac{2P_1}{\mu_1} + \frac{P_1}{\mu_1} + P_2 + (\mu_1 \alpha B K_c)^T (\mu_1 \alpha B K_c)$. Multiplying (19) the left and the right sides by $\begin{bmatrix} I & 0 & 0 \\ 0 & I & -K_c^T \end{bmatrix}$ and its transpose, it is obtained:

$$\begin{bmatrix} Q_1 & \lambda_i P_1 B K_c \\ * & -(1-d_1) P_2 \end{bmatrix} < 0. \tag{20}$$

Note that,

$$\text{He}\{P_1 A + \alpha P_1 B K_c\} + P_2 \leq \text{He}\{P_1 A + \alpha P_1 B K_c\} + P_2 + \alpha^2 (P_1 B K_c)^T (P_1 B K_c)$$

$$= \text{He}\{P_1 A\} + P_2 - \frac{P_1^2}{\mu_1^2} + \left(\frac{P_1}{\mu_1} + \alpha B K_c\right)^T \left(\frac{P_1}{\mu_1} + \alpha B K_c\right), \tag{21}$$

where $\mu_1 > 0$ guarantees that $\alpha B K_c$ would not be too big. The following inequality is introduced:

$$\left(I - \frac{P_1}{\mu_1}\right)\left(I - \frac{P_1}{\mu_1}\right) \geq 0, I - \frac{2P_1}{\mu_1} \geq -\frac{P_1^2}{\mu_1^2}. \tag{22}$$

Combining (22) and (21), it is obtained:

$$\text{He}\{P_1 A + \alpha P_1 B K_c\} + P_2 \leq$$

$$\text{He}\{P_1 A\} + I + P_2 - \frac{2P_1}{\mu_1} + \left(\frac{P_1}{\mu_1} + \alpha B K_c\right)^T \left(\frac{P_1}{\mu_1} + \alpha B K_c\right). \tag{23}$$

Combining (23) and (20), it recovers:

$$\begin{bmatrix} \text{He}\{P_1 A + \alpha P_1 B K_c\} + P_2 & \lambda_i P_1 B K_c \\ * & -(1-d_1) P_2 \end{bmatrix} < 0. \tag{24}$$

The inequality (24) is equal to (18), thus, the linear matrix inequality (10) satisfies (18) and the leader-following formation control problem for the system (1) under communication faults/failures is quadratically stable under (7). Hence, Theorem 1 is proved.

If all conditions hold and the matrices exist, Theorem 1 guarantees a control gain design K_c which is used to achieve and to maintain a desired formation despite communication faults under Assumptions 3. When one or all the agents have communication faults, each faulty agent switches to the control gain K_c reducing the malfunction effects. In the following Section, simulation results are presented to illustrate the strategy effectiveness.

4 Simulation Results: Numerical Examples

A comparison of four simulations is considered, Simulation A1 and Simulation A2 using the classical algorithm (2) with one faulty agent and with faults in the all agents respectively; Simulation B1 and Simulation B2 using our algorithm (7) with one faulty agent and with faults in the all agents respectively. A fleet of six

Table 1 Simulation parameters

Desired positions
$h_1 = [0, 0, 0, 0, 0, 0]^T$
$h_2 = [4, 0, 0, 0, 0, 0]^T$
$h_3 = [6, 2\sqrt{3}, 0, 0, 0, 0]^T$
$h_4 = [4, 4\sqrt{3}, 0, 0, 0, 0]^T$
$h_5 = [0, 4\sqrt{3}, 0, 0, 0, 0]^T$
$h_6 = [-2, 2\sqrt{3}, 0, 0, 0, 0]^T$
Agents' initial positions
$p_1(0) = [0, 0, 0]^T$
$p_2(0) = [1, 3, 0]^T$
$p_3(0) = [-2, -2, 0]^T$
$p_4(0) = [4, -2, 0]^T$
$p_5(0) = [5, 3, 0]^T$
$p_6(0) = [5, 8, 0]^T$

agents ($N = 6$) is considered. In Table 1, the desired positions and the agents' initial positions used in these example are presented.

The agents' initial velocities: $v_i(0) = 0, \forall i = [1, \ldots, 6]$. The target velocity is $v_r = [0, 0, 0.2]^T$. The communication topology is described as follows:

$$\mathcal{L} = \begin{bmatrix} 4 & -1 & -1 & -1 & 0 & -1 \\ -1 & 4 & -1 & -1 & -1 & 0 \\ -1 & -1 & 3 & 0 & 0 & -1 \\ -1 & -1 & 0 & 3 & 0 & -1 \\ 0 & -1 & 0 & 0 & 2 & -1 \\ -1 & 0 & -1 & -1 & -1 & 4 \end{bmatrix}.$$ The communication fault parameters (4) are $\gamma = 0.62$,

$\beta_1 = 1$, and $\beta_2 = 0.6$. For Simulation B1 and B2, the control gain K_c (7) is computed with the decision variables as follows: $\mu_1 = 1$, $\mu_2 = 10$, $d_1 = 0.2$, $\alpha = 1$. All simulations start free fault. After 5 s for simulation A1 and B1, the agent 1 presents communication faults and for simulations A2 and B2, all agents present communication faults.

In Figs. 1 and 2, Definition 1 is illustrated on top and the performance between the velocity of each agent and the target velocity is evaluated using $\|v_i(t) - v_r(t)\|$ on the bottom. When the communication fault occurs, in Simulation A1 and A2, corresponding to Figs. 1a and 2a, the agents lose the formation and the agent velocities become unstable. In Simulation B1 and B2, corresponding to Figs. 1b and 2b, the agents maintain the formation and follow the target velocity.

In Fig. 3, a comparative of derivative of the communication fault is shown. In Fig. 3a, c, the derivative becomes zero because the agents are too far from each other. In Fig. 3b, d, maximum vale of the derivative is less than 1.

In Fig. 4, a comparative of the final position of the agent is shown. In Fig. 4a, c, the formation becomes unstable and the agents cannot maintain it. In Fig. 4b, d, all agents maintain the formation despite the communication fault.

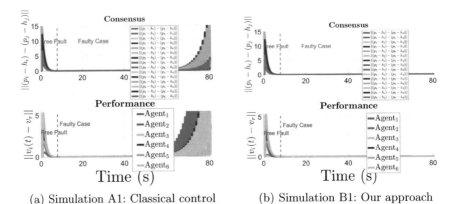

(a) Simulation A1: Classical control   (b) Simulation B1: Our approach

Fig. 1 Comparative of the consensus on top and the performance between each agent velocity and the target velocity on the bottom (Simulation A1 and B1)

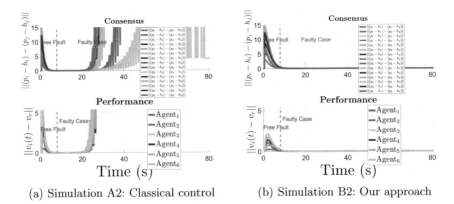

(a) Simulation A2: Classical control   (b) Simulation B2: Our approach

Fig. 2 Comparative of the consensus on top and the performance between each agent velocity and the target velocity on the bottom (Simulation A2 and B2)

In Fig. 5, a comparative of the agent velocities is shown. In Fig. 5a, c, due to the presence of communication failures, velocities are unstable. In Fig. 5b, d, all agent velocities are stable.

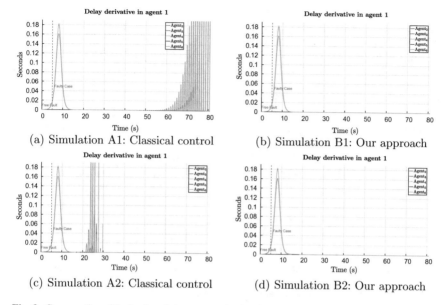

Fig. 3 Comparative of derivative of the communication fault of the agents

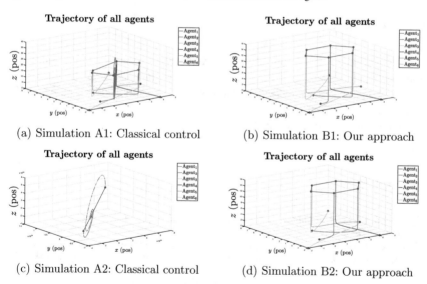

Fig. 4 Comparative of the final formation of the agents

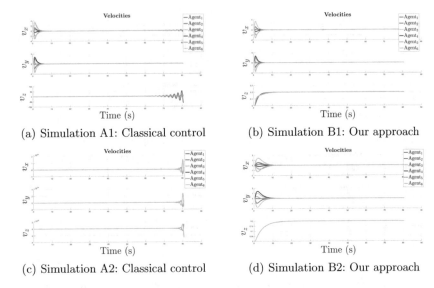

(a) Simulation A1: Classical control   (b) Simulation B1: Our approach

(c) Simulation A2: Classical control   (d) Simulation B2: Our approach

Fig. 5 Comparative of the final formation of the agents

5 Conclusions

A leader-following formation control for multi-agent systems under communication faults/failures has been presented. A control gain has been designed LMI-based conditions that guarantee the stability of the system despite communication faults/failures. The control performance has shown that it can reach and maintain the formation with one or several faulty agents.

References

1. Losada, M.G.: Contributions to Networked and Event-Triggered Control of Linear Systems. Springer (2016)
2. Zhang, Y., Li, R., Zhao, W., Huo, X.: Stochastic leader-following consensus of multi-agent systems with measurement noises and communication time-delays. Neurocomputing **282**, 136–145 (2018)
3. Ren, H., Deng, F.: Mean square consensus of leader-following multi-agent systems with measurement noises and time delays. ISA Trans. **71**, 76 – 83 (2017). Special issue on Distributed Coordination Control for Multi-Agent Systems in Engineering Applications
4. Cui, B., Zhao, C., Ma, T., Feng, C.: Leaderless and leader-following consensus of multi-agent chaotic systems with unknown time delays and switching topologies. Nonlinear Anal. Hybrid Syst. **24**, 115–131 (2017)
5. Lu, M., Liu, L.: Leader-following consensus of multiple uncertain Euler–Lagrange systems subject to communication delays and switching networks. IEEE Trans. Autom. Control **63**(8), 2604–2611 (2017)

6. Wang, X., Su, H.: Self-triggered leader-following consensus of multi-agent systems with input time delay. Neurocomputing **330**, 70–77 (2019)
7. Ye, Y., Su, H.: Leader-following consensus of general linear fractional-order multiagent systems with input delay via event-triggered control. Int. J. Robust Nonlinear Control **28**(18), 5717–5729 (2018)
8. Tan, X.: Leader-following mean square consensus of stochastic multi-agent systems with input delay via event-triggered control. IET Control Theory Appl. **12**(10), 299–309 (2018)
9. Duan, M., Liu, C., Liu, F.: Event-triggered consensus seeking of heterogeneous first-order agents with input delay. IEEE Access **5**, 5215–5223 (2017). https://doi.org/10.1109/ACCESS.2017.2696026
10. Zhao, D., Dong, T.: Reduced-order observer-based consensus for multi-agent systems with time delay and event trigger strategy. IEEE Access **5**, 1263–1271 (2017). https://doi.org/10.1109/ACCESS.2017.2656881
11. Miao, G., Cao, J., Alsaedi, A., Alsaadi, F.E.: Event-triggered containment control for multi-agent systems with constant time delays. J. Franklin Inst. **354**(15), 6956–6977 (2017)
12. Kim, J.H., Park, H.B.: $\mathcal{H}_\infty$ state feedback control for generalized continuous/discrete time-delay system. Automatica **35**(8), 1443–1451 (1999)
13. Langville, A.N., Stewart, W.J.: The Kronecker product and stochastic automata networks. J. Comput. Appl. Math. **167**(2), 429–447 (2004)
14. Zhou, K., Doyle, J.C.: Essentials of Robust Control, vol. 104. Prentice Hall Upper Saddle River, NJ (1998)
15. Olfati-Saber, R., Murray, R.M.: Consensus problems in networks of agents with switching topology and time-delays. IEEE Trans. Autom. Control **49**(9), 1520–1533 (2004)
16. Li, Z., Duan, Z.: Cooperative Control of Multi-Agent Systems: A Consensus Region Approach. CRC Press (2014)
17. Li, Z., Duan, Z., Chen, G.: Dynamic consensus of linear multi-agent systems. IET Control Theory Appl. **5**(9), 19–28 (2011)
18. Georges, J.P., Theilliol, D., Cocquempot, V., Ponsart, J.C., Aubrun, C.: Fault tolerance in networked control systems under intermittent observations. Int. J. Appl. Math. Comput. Sci. **21**(4) (2011)

Fault Diagnosis of Processes and Systems

Regular Approach to Additive Fault Detection in Discrete-Time Linear Descriptor Systems

Dušan Krokavec and Anna Filasová

Abstract The design conditions for regular observer based residual filters for linear discrete-time descriptor systems are presented in the paper, where the problem is stated in combination of linear matrix inequalities and a linear matrix equality. Based on such formulation, they provide a regular interpretation of the problem of designing stable observers for descriptor systems. The results are illustrated by a numerical example in the context of descriptor systems subject to additive fault with projecting the state estimation error to fault residuals.

Keywords Linear discrete-time descriptor systems · Additive faults · Linear observers · Residual filters · Linear matrix inequalities

1 Introduction

Descriptor systems cover a system class appearing in specifical industry fields [5] where the algebraic equations are implicit in inhibiting their direct elimination to obtain dynamic models consisting of pure differential equations. Since the coupled differential and algebraic equations do not constitute a standard state-space description [8], and because these two specificities can be considered separately, comprehensive frameworks for the analysis and synthesis are proposed in [1, 11, 12], giving a reader also the real practice state. However, focusing on fault detection and fault tolerant control problems, the possible restriction of these results have to be considered.

D. Krokavec (✉) · A. Filasová
Faculty of Electrical Engineering and Informatics, Department of Cybernetics and Artificial Intelligence, Technical University of Košice, Letná, 04200 Košice, Slovakia
e-mail: dusan.krokavec@tuke.sk

A. Filasová
e-mail: anna.filasova@tuke.sk

© The Author(s), under exclusive license to Springer Nature Switzerland AG 2021
J. Korbicz et al. (eds.), *Advances in Diagnostics of Processes and Systems*,
Studies in Systems, Decision and Control 313,
https://doi.org/10.1007/978-3-030-58964-6_5

Limiting in fault detection and estimation, adaptive observer principles are preferred for this system class in relation to additive faults (see, e.g., [3, 6, 7]), also comprehending substantial modifications with respect to discrete-time descriptor system models [9, 10]. The natural drawback of these standard methods is that the associated state observer also have a descriptor structure.

Adapting an approach formulated in [2] in this paper is presented a procedure synthesizing regular state observers and residual filters for linear discrete-time descriptor systems, subjected to additive faults, restricted by system equivalency constraints and limited by performance requirements expressed in the form of an H_∞ norm disturbance attenuation. In view of the above features, the proposed method is based on using the Lyapunov function to enforce (possibly competing) design specifications. Since H_∞ norm approach allows to impose directly several constraints formulated as a single criterion reflecting all specifications, this formulation is preferred. In an linear matrix inequality (LMI) based design, a set of LMI is used to cover stability and disturbance attenuation constraints, and linear matrix equality (LME) is reflecting specification implying from restricted system equivalency. The approach thus reduces the synthesis problem to a feasible solving of the defined regular observer parameters.

The paper is organized as follows. Ensuing introduction in Sect. 1, state estimation principles for discrete-time linear descriptor systems are presented in Sect. 2, highlighting the methodology introduced in [2] and its generalization in [1, 10]. To eliminate constraints implying from singular value decomposition (SVD), or others potentially generated by matrix pseudo-inverse procedures, design conditions for regular state observers, reflecting optimization based on linear matrix structures formalisation, are presented in Sect. 3. On the other hand, reformulating the same stability problems, design procedure for regular discrete-time residual filters, associated to linear descriptor systems, are provided in Sect. 4. The applicability of these results by means of two specific examples is illustrated in Sects. 5 and 6 draws some conclusions.

Throughout the paper the following notation is used: $X < 0$ conveys that a square matrix X is symmetric and negative definite, $\rho(X)$ reckons eigenvalue spectrum of the square matrix X, x^T, X^T denote the transpose of the vector x and the matrix X, the symbol I_n indicates the n-th order unit matrix, diag[$\cdot$] enters up a (block) diagonal matrix, $\mathbb{R}$ denotes the set of real numbers, and $\mathbb{R}^n$, $\mathbb{R}^{n \times r}$ refer to the set of all n-dimensional real vectors and $n \times r$ real matrices, respectively.

2   System Model Properties

A linear descriptor discrete-time multi input, multi output (MIMO) system in presence of an unknown additive fault is described by the state-space equations

$$Eq(i + 1) = Fq(i) + Gu(i) + Hf(i) + Dd(i), \tag{1}$$

$$y(i) = Cq(i) + o(i),\qquad(2)$$

where $q(i) \in \mathbb{R}^n$, $u(i) \in \mathbb{R}^r$, $y(i) \in \mathbb{R}^m$, $f(i) \in \mathbb{R}^s$ are the state, the input, the measurement output and additive fault vector, respectively, $d(i) \in \mathbb{R}^p$ is the unknown disturbances vector and $v(i) \in \mathbb{R}^m$ is the measurement noise vector, $F \in \mathbb{R}^{n \times n}$, $G \in \mathbb{R}^{n \times r}$, $C \in \mathbb{R}^{p \times n}$, $D \in \mathbb{R}^{n \times p}$ are known system matrices, $H \in \mathbb{R}^{n \times s}$ is the known fault distribution matrix. It is supposed that $E \in \mathbb{R}^{n \times n}$ is singular matrix such that

$$\operatorname{rank} \begin{bmatrix} E \\ C \end{bmatrix} = n, \quad \operatorname{rank} E < m < n.\qquad(3)$$

In order to reformulate a descriptor approach in observer synthesis the following can be recalled.

Proposition 1 ([4] system pair admissibility) *Considering an autonomous linear descriptor system model of the form*

$$Eq(i+1) = Fq(i),\qquad(4)$$

then the pair (E, F) is regular, causal and asymptotically stable if and only if there exists a regular symmetric matrix Z of appropriate dimension that

$$F^{\mathrm{T}}ZF - E^{\mathrm{T}}ZE < 0,\qquad(5)$$

$$E^{\mathrm{T}}Z = ZE \geq 0.\qquad(6)$$

This proposition, based on the generalized Lyapunov inequality, leads to the use of the descriptor form of the state observer and is therefore mainly used to analyze the stability of discrete-time linear descriptor systems.

The regular form of the linear state observer for discrete-time linear descriptor systems can be derived from the restricted system equivalency principle [1]. Using the singular value decomposition (SVD) the following proposition yields.

Proposition 2 ([2]) *If (3) is satisfied then there exists such a matrix $\Lambda \in \mathbb{R}^{(n+m) \times n}$ that the relation*

$$\Lambda \begin{bmatrix} E \\ C \end{bmatrix} = \begin{bmatrix} T & N \\ O & M \end{bmatrix} \begin{bmatrix} E \\ C \end{bmatrix}\qquad(7)$$

implies

$$TE + NC = I_n, \quad OE + MC = 0,\qquad(8)$$

where $T \in \mathbb{R}^{n \times n}$, $N \in \mathbb{R}^{n \times m}$, $O \in \mathbb{R}^{m \times n}$, $M \in \mathbb{R}^{m \times m}$.

Corollary 1 *Using SVD then if (3) is satisfied it yields*

$$X \begin{bmatrix} E \\ C \end{bmatrix} = \begin{bmatrix} X_{11} & X_{12} \\ X_{21} & X_{22} \end{bmatrix} \begin{bmatrix} E \\ C \end{bmatrix} = \begin{bmatrix} \Xi \\ 0 \end{bmatrix} Y = \begin{bmatrix} \Xi Y \\ 0 \end{bmatrix},\qquad(9)$$

where with $X \in \mathbb{R}^{(n+m)\times(n+m)}$, $Y, \Xi \in \mathbb{R}^{n\times n}$, $h = 1, 2 \ldots, n - 1$,

$$XX^{\mathrm{T}} = I_n, \quad YY^{\mathrm{T}} = I_m, \quad \Xi = diag\left[\sigma_1\ \sigma_2\ \cdots\sigma_n\right], \quad \sigma_h \geq \sigma_{h+1}. \tag{10}$$

Thus, evidently,

$$(\Xi Y)^{-1}\left[X_{11}\ X_{12}\right]\begin{bmatrix} E \\ C \end{bmatrix} = I_n, \quad \left[X_{21}\ X_{22}\right]\begin{bmatrix} E \\ C \end{bmatrix} = 0, \tag{11}$$

$$T = (\Xi Y)^{-1}X_{11}, \quad N = (\Xi T)^{-1}X_{12}, \quad O = X_{21}, \quad M = X_{22}. \tag{12}$$

The structure of the observer for faulty-free system can be defined as [2]

$$p(i + 1) = Lp(i) + Ku(i) + (J_1 + J_2)y(i), \tag{13}$$

$$q_e(i) = p(i) + Ny(i) + J_3My(i), \tag{14}$$

where $L \in \mathbb{R}^{n\times n}$, $J_1, J_2, J_2 \in \mathbb{R}^{n\times m}$ and $K \in \mathbb{R}^{n\times r}$ must be determined in such a way that $q_e(i)$ asymptotically converges to $q(i)$.

3 Regular State Observer Design

To eliminate SVD or matrix pseudo-inverse procedure, and to release the design constraints, it is possible to formulate the design task taking into account only the condition [10]

$$TE + NC = I_n \tag{15}$$

and defining the structure of the observer for faulty-free system

$$p(i + 1) = TFp(i) + TGu(i) + J(y(i) - Cq_e(i)), \tag{16}$$

$$q_e(i) = p(i) + Ny(i). \tag{17}$$

The presented design methodology goes in this direction but considering $T \in \mathbb{R}^{n\times n}$ and $N \in \mathbb{R}^{n\times m}$ as matrix variables to define a set of LMIs which guaranties, if the task is feasible, the matrix TF is Schur and a matrix equality related to (15) is satisfied.

Theorem 1 *The discrete-time state estimation structure (16), (17) related to discrete-time linear descriptor fault-free system (1), (2) is stable if there exist a symmetric positive definite matrix $P \in \mathbb{R}^{n\times n}$, matrices $V \in \mathbb{R}^{n\times n}$, $Q, S \in \mathbb{R}^{n\times m}$ and a positive scalar $\gamma \in \mathbb{R}$ such that*

$$P = P^{\mathrm{T}} \succ 0, \quad \gamma > 0, \tag{18}$$

$$\begin{bmatrix} -P & * & * & * & * \\ 0 & -\gamma I_p & * & * & * \\ 0 & 0 & -\gamma I_m & * & * \\ VF - QC & VD & -Q & -P & * \\ C & 0 & 0 & 0 & -\gamma I_m \end{bmatrix} \prec 0, \qquad (19)$$

$$VE + SC = P. \qquad (20)$$

When the above conditions hold

$$J = P^{-1}Q, \quad T = P^{-1}V, \quad N = P^{-1}S. \qquad (21)$$

*Hereafter, * denotes the symmetric item in a symmetric matrix.*

Proof With the constraint (15) one can see that at the time instant $i + 1$ it has to be

$$\begin{aligned} q(i+1) &= (TE + NC)q(i+1) \\ &= T(Fq(i) + Gu(i) + Dd(i)) + Ny(i+1). \end{aligned} \qquad (22)$$

Reflecting the system state error defined by

$$e(i) = q(i) - q_e(i), \qquad (23)$$

then substituting for $q(i+1)$, $q_e(i+1)$ it yields

$$\begin{aligned} e(i+1) &= T(Fq(i) + Gu(i) + Dd(i)) + Ny(i+1) - \\ &\quad - TFp(i) - TGu(i) - J(y(i) - Cq_e(i)) - Ny(i+1) \\ &= (TF - JC)e(i) + TDd(i)) - Jo(i) \qquad (24) \\ &= (TF - JC)e(i) + \begin{bmatrix} TD & -J \end{bmatrix} \begin{bmatrix} d(i) \\ o(i) \end{bmatrix}. \end{aligned}$$

Considering the parameters involved in the descriptor form (1), (2) and a benefit of the H_∞ norm constraint, the Lyapunov function is constructed as

$$v(e(i))e^{\mathrm{T}}(i)Pe(i) + \gamma^{-1}\sum_{j=0}^{i-1}(e_y^{\mathrm{T}}(j)e_y(j) - \gamma^2 d^{\circ \mathrm{T}}(j)d^{\circ}(j)) > 0, \qquad (25)$$

where

$$e_y(i) = Ce(i), \quad d^{\circ}(i) = \begin{bmatrix} d(i) \\ o(i) \end{bmatrix}, \quad D^{\circ} = \begin{bmatrix} TD & -J \end{bmatrix}, \quad F_e = TF - JC \qquad (26)$$

and it has to yield for all observer error trajectories

$$\Delta v(e(i)) = e^{\mathrm{T}}(i+1)Pe(i+1) - e^{\mathrm{T}}(i)Pe(i) + \\ + \gamma^{-1}e_y^{\mathrm{T}}(i)e_y(i) - \gamma d^{\circ\mathrm{T}}(i)d^{\circ}(i) < 0. \tag{27}$$

Then, combining vector variables into the composed vector

$$e^{\circ\mathrm{T}}(i) = \left[e^{\mathrm{T}}(i) \; d^{\circ\mathrm{T}}(i) \right], \tag{28}$$

it can state the following

$$\Delta v(e^{\circ}(i)) = e^{\circ\mathrm{T}}(i)P^{\circ}e^{\circ}(i) < 0, \tag{29}$$

where

$$P^{\circ} = \begin{bmatrix} -P + \gamma^{-1}C^{\mathrm{T}}C + F_e^{\mathrm{T}}PF_e & F_e^{\mathrm{T}}PD^{\circ} \\ D^{\circ\mathrm{T}}PF_e & -\gamma I_{p+m} + D^{\circ\mathrm{T}}PD^{\circ} \end{bmatrix} \prec 0. \tag{30}$$

Therefore, partitioning the matrix inequality (30) in accord to the dimensions of its blocks

$$\begin{bmatrix} -P + \gamma^{-1}C^{\mathrm{T}}C & 0 \\ 0 & -\gamma I_{p+m} \end{bmatrix} + \begin{bmatrix} F_e^{\mathrm{T}} \\ D^{\circ\mathrm{T}} \end{bmatrix} P \left[F_e \; D^{\circ} \right] \prec 0, \tag{31}$$

(31) is rearranged as

$$\begin{bmatrix} -P + \gamma^{-1}C^{\mathrm{T}}C & 0 & F_e^{\mathrm{T}}P \\ 0 & -\gamma I_{p+m} & D^{\circ\mathrm{T}}P \\ PF_e & PD^{\circ} & -P \end{bmatrix} \prec 0, \tag{32}$$

$$\begin{bmatrix} -P & 0 & F_e^{\mathrm{T}}P & C^{\mathrm{T}} \\ 0 & -\gamma I_{p+m} & D^{\circ\mathrm{T}}P & 0 \\ PF_e & PD^{\circ} & -P & 0 \\ C & 0 & 0 & -\gamma I_m \end{bmatrix} \prec 0, \tag{33}$$

respectively, applying the Schur complement property.

At this point it is easy to see that

$$PF_e = P(TF - JC) = VF - QC, \tag{34}$$

$$PD^{\circ} = P\left[TD - J \right] = \left[VD - Q \right], \tag{35}$$

where

$$V = PT, \quad Q = PJ. \tag{36}$$

Since

$$I_{p+m} = \begin{bmatrix} I_p & \\ & I_m \end{bmatrix},$$ (37)

then (33) can be rewritten as (19).

Finally, multiplying the left side of (15) by P it yields

$$P(TE + NC) = VE + SC = P,$$ (38)

where

$$S = PN.$$ (39)

This equality is equivalent to (15) and gives (20). This concludes the proof. ■

The coupling between system matrices and the Lyapunov matrix P in (20) causes the result is quite conservative in general. Using an additive slack matrix this issue is relaxed. The principle is tightly related to the following theorem.

Theorem 2 *The discrete-time state estimation structure (16), (17) related to discrete-time linear descriptor fault-free system (1), (2) is stable if there exist symmetric positive definite matrices $P, U \in \mathbb{R}^{n \times n}$, matrices $V \in \mathbb{R}^{n \times n}$, $Q, S \in \mathbb{R}^{n \times m}$ and a positive scalar $\gamma \in \mathbb{R}$ such that*

$$P = P^{\mathrm{T}} \succ 0, \quad U = U^{\mathrm{T}} \succ 0, \quad \gamma > 0,$$ (40)

$$\begin{bmatrix} -P & * & * & * & * \\ 0 & -\gamma I_p & * & * & * \\ 0 & 0 & -\gamma I_m & * & * \\ VF - QC & VD & -Q & -2U + P & * \\ C & 0 & 0 & 0 & -\gamma I_m \end{bmatrix} \prec 0,$$ (41)

$$VE + SC = U.$$ (42)

When the above conditions hold

$$J = U^{-1}Q, \quad T = U^{-1}V, \quad N = U^{-1}S.$$ (43)

Proof Applying relations (24), (26) it yields that

$$F_e e(i) + D^\circ d^\circ(i) - e(i+1) = 0$$ (44)

and if $U \in \mathbb{R}^{n \times n}$ is a symmetric positive definite matrix it is satisfied

$$e(i+1)^{\mathrm{T}} U (F_e e(i) + D^\circ d^\circ(i) - e(i+1)) = 0.$$ (45)

Adding (45) and its transpose to (27) results the relation

$$
\begin{aligned}
\Delta v(e(i)) = {} & e^{\mathrm{T}}(i+1) \boldsymbol{P} e(i+1) - e^{\mathrm{T}}(i) \boldsymbol{P} e(i) + \\
& + \gamma^{-1} e_y^{\mathrm{T}}(i) e_y(i) - \gamma d^{\circ \mathrm{T}}(i) d^{\circ}(i) + \\
& + e(i+1)^{\mathrm{T}} \boldsymbol{U} (\boldsymbol{F}_e e(i) + \boldsymbol{D}^{\circ} d^{\circ}(i) - e(i+1)) + \\
& + (\boldsymbol{F}_e e(i) + \boldsymbol{D}^{\circ} d^{\circ}(i) - e(i+1))^{\mathrm{T}} \boldsymbol{U} e(i+1) \\
& < 0
\end{aligned}
\tag{46}
$$

and combining vector variables into the expanded composed vector

$$
e^{\bullet \mathrm{T}}(i) = \begin{bmatrix} e^{\mathrm{T}}(i) \; d^{\circ \mathrm{T}}(i) \; e(i+1)^{\mathrm{T}} \end{bmatrix},
\tag{47}
$$

the condition which defines stability problem is characterized as

$$
\Delta v(e^{\bullet}(i)) = e^{\bullet \mathrm{T}}(i) \boldsymbol{P}^{\bullet} e^{\bullet}(i) < 0,
\tag{48}
$$

where

$$
\boldsymbol{P}^{\bullet} = \begin{bmatrix} -\boldsymbol{P} + \gamma^{-1} \boldsymbol{C}^{\mathrm{T}} \boldsymbol{C} & \boldsymbol{0} & \boldsymbol{F}_e^{\mathrm{T}} \boldsymbol{P} \\ \boldsymbol{0} & -\gamma \boldsymbol{I}_{p+m} & \boldsymbol{D}^{\circ \mathrm{T}} \boldsymbol{U} \\ \boldsymbol{U} \boldsymbol{F}_e & \boldsymbol{P} \boldsymbol{D}^{\circ} & -2\boldsymbol{U} + \boldsymbol{P} \end{bmatrix} \prec 0.
\tag{49}
$$

Naturally, it is possible to construct a more general structure of (49)

$$
\begin{bmatrix} -\boldsymbol{P} & \boldsymbol{0} & \boldsymbol{F}_e^{\mathrm{T}} \boldsymbol{P} & \boldsymbol{C}^{\mathrm{T}} \\ \boldsymbol{0} & -\gamma \boldsymbol{I}_{p+m} & \boldsymbol{D}^{\circ \mathrm{T}} \boldsymbol{U} & \boldsymbol{0} \\ \boldsymbol{U} \boldsymbol{F}_e & \boldsymbol{P} \boldsymbol{D}^{\circ} & -2\boldsymbol{U} + \boldsymbol{P} & \boldsymbol{0} \\ \boldsymbol{C} & \boldsymbol{0} & \boldsymbol{0} & -\gamma \boldsymbol{I}_m \end{bmatrix} \prec 0.
\tag{50}
$$

Since it can write immediately

$$
\boldsymbol{U} \boldsymbol{F}_e = \boldsymbol{U}(\boldsymbol{T} \boldsymbol{F} - \boldsymbol{J} \boldsymbol{C}) = \boldsymbol{V} \boldsymbol{F} - \boldsymbol{Q} \boldsymbol{C},
\tag{51}
$$

$$
\boldsymbol{U} \boldsymbol{D}^{\circ} = \boldsymbol{U} \begin{bmatrix} \boldsymbol{T} \boldsymbol{D} & -\boldsymbol{J} \end{bmatrix} = \begin{bmatrix} \boldsymbol{V} \boldsymbol{D} & -\boldsymbol{Q} \end{bmatrix},
\tag{52}
$$

where now

$$
\boldsymbol{V} = \boldsymbol{U} \boldsymbol{T}, \quad \boldsymbol{Q} = \boldsymbol{U} \boldsymbol{J},
\tag{53}
$$

then, using (37) inequality (50) implies (41).
Multiplying the left side of (15) by $\boldsymbol{U}$ the condition becomes

$$
\boldsymbol{U}(\boldsymbol{T} \boldsymbol{E} + \boldsymbol{N} \boldsymbol{C}) = \boldsymbol{V} \boldsymbol{E} + \boldsymbol{S} \boldsymbol{C} = \boldsymbol{U}, \quad \boldsymbol{S} = \boldsymbol{U} \boldsymbol{N}.
\tag{54}
$$

This equality is equivalent to (15) and gives (42). This concludes the proof. ∎

Notice that the above descriptions of observer asymptotic stability in terms of linear matrix inequalities and linear matrix equality give the possibility to request that the observer matrix parameters have given particular structures.

4 Regular Residual Filter Design

To give answers regarding action affects of unknown amplitudes of additive faults and disturbances on the fault residuals, the H_∞ norm bounds δ, γ of associated transfer functions are proposed to be interactively optimized using the following theorems.

Theorem 3 *The discrete-time residual filter built on the state estimation structure (16), (17) related to discrete-time linear descriptor fault-free system (1), (2) is stable if there exist a symmetric positive definite matrix $P \in \mathbb{R}^{n \times n}$, matrices $V \in \mathbb{R}^{n \times n}$, $Q, S \in \mathbb{R}^{n \times m}$, $R \in \mathbb{R}^{m \times m}$ and positive scalars $\gamma, \delta \in \mathbb{R}$ such that*

$$P = P^T \succ 0, \quad \gamma > 0, \quad \delta > 0, \tag{55}$$

$$\begin{bmatrix} -P & * & * & * & * & * & * \\ 0 & -\gamma I_p & * & * & * & * & * \\ 0 & 0 & -\gamma I_m & * & * & * & * \\ 0 & 0 & 0 & -\delta I_s & * & * & * \\ VF - QC & VD & -Q & VH & -P & * & * \\ RC & 0 & 0 & 0 & 0 & -\gamma I_m & * \\ RC & 0 & 0 & 0 & 0 & 0 & -\delta I_m \end{bmatrix} \prec 0, \tag{56}$$

$$VE + SC = P. \tag{57}$$

When the above conditions hold the observer related residual filter parameters are given by (21) and residual gain is defined by the matrix variable R.

Proof Applying the filter gain matrix $R \in \mathbb{R}^{m \times m}$ to define residuals as

$$r(i) = RCe(i), \tag{58}$$

then (1), (2) implies

$$\begin{aligned} e_f(i+1) &= e(i+1) + THf(i) \\ &= F_e e(i) + D^\circ d^\circ(i) + THf(i), \end{aligned} \tag{59}$$

where $e(i+1)$ is introduced in (24), (26). Defining the Lyapunov function as

$$v(e_f(i)) = e_f^T(i) P e_f(i) + \delta^{-1} \sum_{j=0}^{i-1} (r^T(j)r(j) - \delta^2 f^T(j)f(j))$$

$$+ \gamma^{-1} \sum_{j=0}^{i-1} (r^T(j)r(j) - \gamma^2 d^{\circ T}(j)d^{\circ}(j)) \quad (60)$$

$$> 0,$$

it has to yield

$$\Delta v(e_f(i)) = e_f^T(i+1) P e_f(i+1) - e_f^T(i) P e_f(i) +$$
$$+ \delta^{-1} r^T(i)r(i) - \delta f^T(i)f(i) + \gamma^{-1} r^T(i)r(i) - \gamma d^{\circ T}(i)d^{\circ}(i) \quad (61)$$
$$< 0.$$

Then, combining vector variables into the composed vector

$$e^{\circ T}(i) = \left[e^{\circ T}(i) \ f^T(i) \right] = \left[e^T(i) \ d^{\circ T}(i) \ f^T(i) \right], \quad (62)$$

which, by solving for $\Delta v(e_f(i))$, gives

$$\Delta v(e^{\circ}(i)) = e^{\circ T}(i) P^{\circ} e^{\circ}(i) < 0, \quad (63)$$

and with $H^{\circ} = T H$ it follows that

$$P^{\circ} = \begin{bmatrix} P_{11}^{\circ} & F_e^T P D^{\circ} & F_e^T P H^{\circ} \\ D^{\circ T} P F_e & -\gamma I_{p+m} + D^{\circ T} P D^{\circ} & D^{\circ T} P H^{\circ} \\ H^{\circ T} P F_e & H^{\circ T} P D^{\circ} & -\delta I_s + H^{\circ T} P H^{\circ} \end{bmatrix} \prec 0, \quad (64)$$

$$P_{11}^{\circ} = -P + (\gamma^{-1} + \delta^{-1}) C^T R^T R C + F_e^T P F_e. \quad (65)$$

Analogously with (31)–(33) it can arrive to the inequality

$$\begin{bmatrix} -P & 0 & 0 & F_e^T P & C^T R^T & C^T R^T \\ 0 & -\gamma I_{p+m} & 0 & D^{\circ T} P & 0 & 0 \\ 0 & 0 & -\delta I_s & H^{\circ T} P & 0 & 0 \\ P F_e & P D^{\circ} & P H^{\circ} & -P & 0 & 0 \\ R C & 0 & 0 & 0 & -\gamma I_m & 0 \\ R C & 0 & 0 & 0 & 0 & -\delta I_m \end{bmatrix} \prec 0 \quad (66)$$

and reflecting (34)–(36) then (66) implies (56). Finally, (20) is copied into (57). This concludes the proof. ∎

Theorem 4 *The discrete-time residual filter built on the state estimation structure (16), (17) related to discrete-time linear descriptor fault-free system (1), (2) is stable if there exist symmetric positive definite matrices $P, U \in \mathbb{R}^{n \times n}$, matrices $V \in \mathbb{R}^{n \times n}$, $Q, S \in \mathbb{R}^{n \times m}$, $R \in \mathbb{R}^{m \times m}$ and positive scalars $\gamma, \delta \in \mathbb{R}$ such that*

$$P = P^{\mathrm{T}} \succ 0, \quad U = U^{\mathrm{T}} \succ 0, \quad \gamma > 0, \quad \delta > 0, \tag{67}$$

$$\begin{bmatrix} -P & * & * & * & * & * & * \\ 0 & -\gamma I_p & * & * & * & * & * \\ 0 & 0 & -\gamma I_m & * & * & * & * \\ 0 & 0 & 0 & -\delta I_s & * & * & * \\ VF - QC & VD & -Q & VH & -2U + P & * & * \\ RC & 0 & 0 & 0 & 0 & -\gamma I_m & * \\ RC & 0 & 0 & 0 & 0 & 0 & -\delta I_m \end{bmatrix} \prec 0, \tag{68}$$

$$VE + SC = U. \tag{69}$$

*When the above conditions hold the observer related residual filter parameters are given by (43) and residual gain is defined by the matrix variable **R**.*

The proof follows immediately from Theorem 2 by observing the expanded relation (59).

5 Illustrative Example

In the example, the input-output dynamics of the system (1), (2) is given by the state-space representation with the parameters $T_s = 0.1$ s,

$$F = \begin{bmatrix} 1.1782 & 0.0014 & 0.5116 & -0.4033 \\ -0.0514 & 0.6619 & -0.0110 & 0.0613 \\ 0.0762 & 0.3351 & 0.5606 & 0.3824 \\ -0.0006 & 0.3353 & 0.0893 & 0.8494 \end{bmatrix}, \quad G = H = \begin{bmatrix} 0.0045 & -0.0876 \\ 0.4672 & 0.0012 \\ 0.2132 & -0.2353 \\ 0.2131 & -0.0161 \end{bmatrix},$$

$$E = \begin{bmatrix} 1 & \\ 0 & \\ & 1 \\ & 1 \end{bmatrix}, \quad D = \begin{bmatrix} -0.0831 \\ 0.4684 \\ -0.0221 \\ 0.1970 \end{bmatrix}, \quad C = \begin{bmatrix} 1 & 0 & 1 & 0 \\ 0 & 1 & 0 & 1 \end{bmatrix}.$$

Solving (18)–(20) using SeDuMi package, the design task is feasible and the LMI matrix variables are

$$P = \begin{bmatrix} 3.8924 & -0.2960 & 0.4001 & 0.2054 \\ -0.2960 & 3.4207 & 0.3078 & 0.8800 \\ 0.4001 & 0.3078 & 4.0455 & -0.0946 \\ 0.2054 & 0.8800 & -0.0946 & 4.0590 \end{bmatrix}, \quad Q = \begin{bmatrix} 0.1696 & -0.5142 \\ -0.2492 & -0.9305 \\ 0.2162 & 0.6102 \\ 0.2850 & 1.2084 \end{bmatrix},$$

$$V = \begin{bmatrix} 1.1384 & -0.0367 & -2.3539 & 0.5014 \\ -0.3799 & -0.0050 & 0.2239 & -2.5407 \\ -0.6906 & -0.0532 & 2.9548 & -0.4024 \\ 0.3344 & 0.0153 & 0.0344 & 3.1790 \end{bmatrix}, \quad S = \begin{bmatrix} 0.1114 & 0.0401 \\ 0.0389 & 0.4222 \\ 0.0585 & 0.2194 \\ 0.0029 & -0.3076 \\ 0.0128 & 0.3989 \\ -0.0981 & 0.1311 \end{bmatrix},$$

$$\gamma = 5.1303 \,.$$

Immediately, the discrete-time observer gain matrices are computed directly as

$$J = \begin{bmatrix} 0.0249 & -0.2065 \\ -0.0998 & -0.4125 \\ 0.0607 & 0.2121 \\ 0.0920 & 0.4025 \end{bmatrix},$$

$$T = \begin{bmatrix} 0.3011 & -0.0086 & -0.6989 & 0 \\ -0.0888 & -0.0023 & -0.0888 & -1 \\ -0.1918 & -0.0120 & 0.8082 & 0 \\ 0.0819 & 0.0044 & 0.0819 & 1 \end{bmatrix}, \quad N = \begin{bmatrix} 0.6989 & 0 \\ 0.0888 & 1 \\ 0.1918 & 0 \\ -0.0819 & 0 \end{bmatrix}.$$

The obtained observer parameters force the stable discrete-time observer, where

$$F_e = \begin{bmatrix} 0.2771 & -0.0329 & -0.2626 & -0.1827 \\ -0.0109 & 0.0459 & -0.0847 & -0.4352 \\ -0.2246 & 0.0505 & 0.2943 & 0.1736 \\ 0.0099 & -0.0368 & 0.0851 & 0.4455 \end{bmatrix}, \quad \rho(F_e) = \begin{Bmatrix} 0.6072 \\ 0.4015 \\ 0.0448 \\ 0.0093 \end{Bmatrix},$$

$$TF = \begin{bmatrix} 0.3020 & -0.2394 & -0.2377 & -0.3892 \\ -0.1107 & -0.3666 & -0.1845 & -0.8477 \\ -0.1638 & 0.2626 & 0.3551 & 0.3856 \\ 0.1020 & 0.3658 & 0.1771 & 0.8480 \end{bmatrix}, \quad \rho(TF) = \begin{Bmatrix} 0.6030 \\ 0.4005 \\ 0.1341 \\ 0.0009 \end{Bmatrix}.$$

Solving (67)–(69) with optimized $\delta = 18.1$, $R = 0.16I_2$, the LMI variables are

$$P = \begin{bmatrix} 13.0139 & -1.2534 & 1.0589 & 0.5765 \\ -1.2534 & 11.8907 & 1.3342 & 3.0965 \\ 1.0589 & 1.3342 & 13.2491 & 0.0045 \\ 0.5765 & 3.0965 & 0.0045 & 14.7848 \end{bmatrix}, \quad Q = \begin{bmatrix} 0.6746 & -1.5638 \\ -0.7444 & -3.1038 \\ 0.4191 & 1.7337 \\ 0.9116 & 3.8706 \end{bmatrix},$$

$$V = \begin{bmatrix} 4.1945 & 0.3744 & -8.0140 & 1.4929 \\ -1.0647 & 1.3420 & 0.7215 & -9.6375 \\ -2.8525 & -0.7107 & 9.7046 & -1.1854 \\ 0.9893 & -1.1687 & 0.0972 & 11.6870 \end{bmatrix}, \quad S = \begin{bmatrix} 8.9685 & -0.9003 \\ 0.1644 & 11.7633 \\ 3.8070 & 0.8859 \\ -0.3967 & 2.1258 \end{bmatrix},$$

$$U = \begin{bmatrix} 13.1630 & -0.9003 & 0.9545 & 0.5926 \\ -0.9003 & 11.7633 & 0.8859 & 2.1258 \\ 0.9545 & 0.8859 & 13.5116 & -0.2995 \\ 0.5926 & 2.1258 & -0.2995 & 13.8128 \end{bmatrix}, \quad \gamma = 17.2940$$

and the discrete-time fault observer gain matrices are

$$J = \begin{bmatrix} 0.0400 & -0.1709 \\ -0.0767 & -0.3523 \\ 0.0349 & 0.1711 \\ 0.0768 & 0.3455 \end{bmatrix},$$

$$T = \begin{bmatrix} 0.3284 & 0.0481 & -0.6716 & 0 \\ -0.0593 & 0.1428 & -0.0593 & -1 \\ -0.2291 & -0.0678 & 0.7709 & 0 \\ 0.0617 & -0.1101 & 0.0617 & 1 \end{bmatrix}, \quad N = \begin{bmatrix} 0.6716 & 0 \\ 0.0593 & 1 \\ 0.2291 & 0 \\ -0.0617 & 0 \end{bmatrix}.$$

The obtained system matrices are stable since

$$F_e = \begin{bmatrix} 0.2933 & -0.0219 & -0.2490 & -0.2154 \\ -0.0043 & 0.0915 & -0.0777 & -0.4872 \\ -0.2426 & 0.0420 & 0.2808 & 0.2119 \\ 0.0056 & -0.0623 & 0.0798 & 0.4959 \end{bmatrix}, \quad \rho(F_e) = \begin{Bmatrix} 0.6612 \\ 0.4312 \\ 0.0461 \\ 0.0231 \end{Bmatrix},$$

$$TF = \begin{bmatrix} 0.3333 & -0.1927 & -0.2090 & -0.3863 \\ -0.0811 & -0.2607 & -0.1544 & -0.8394 \\ -0.2077 & 0.2131 & 0.3158 & 0.3830 \\ 0.0824 & 0.2831 & 0.1566 & 0.8414 \end{bmatrix}, \quad \rho(TF) = \begin{Bmatrix} 0.6496 \\ 0.4263 \\ 0.1241 \\ 0.0298 \end{Bmatrix}.$$

These results characterize the discrete-time state observer and discrete-time residual filter performances prescribed by the synthesis strategy. Due to limited scope of the contribution, additional solutions are not presented.

6 Concluding Remarks

This paper investigates the fault detection problem for discrete-time linear descriptor system. In this regards, new theorems are introduced to design a regular system state observer and to synthesis residual filters built on a such observer structure. The explicit expression of the filter gain matrices is characterized in terms of solution to a set of LMI and LME. The conditions improve the design consistency, guarantee the asymptotic stability of residual filters and minimize the impact of disturbance on residual signals.

Acknowledgements The work presented was supported by VEGA, the Grant Agency of the Ministry of Education and the Academy of Science of Slovak Republic under Grant No. 1/0608/17. This support is gratefully acknowledged.

References

1. Dai, L.: Singular Control Systems. Springer, Berlin (1989). https://doi.org/10.1007/BFb000. 2475
2. Darouach, M., Boutayeb, M.: Design of observers for descriptor systems. IEEE Trans. Autom. Control **40**(7), 1323–1321 (1995). https://doi.org/10.1109/9.400.467
3. Gao, Z., Ding, S.X.: Fault estimation and fault-tolerant control for descriptor systems via proportional, multiple-integral and derivative observer design. IET Control Theory Appl. **1**(5), 1208–1218 (2007). https://doi.org/10.1049/iet-cta:20060389
4. Hsiung, K.L., Lee, L.: Lyapunov inequality and bounded real lemma for discrete-time descriptor systems. IEE Proc. Control Theory Appl. **146**(4), 327–331 (1999). https://doi.org/10.1049/ip-cta:19990451
5. Kumar, A., Daoutidis, P.: Control of Nonlinear Differential Algebraic Equation Systems with Applications to Chemical Processes. Chapman and Hall, Boca Raton (1999)
6. Ren, J., Zhang, Q.: PD observer design for descriptor systems. An LMI approach. Int. J. Control Autom. Syst. **8**(4), 735–740 (2010). https://doi.org/10.1007/s12555-010-0404-4
7. Shi, F., Patton, R.J.: Simultaneous state and fault estimation for descriptor systems using an augmented PD observer. IFAC Proc. Vol. **47**(3), 8006–8011 (2014). https://doi.org/10.3182/20140824-6-ZA-1003.01383
8. Verghese, G.C., Levy, B.C., Kailath, T.L: A generalized state-space for singular systems. IEEE Trans. Autom. Control **26**(4), 811–831 (1981). https://doi.org/10.1109/TAC.1981.1102763
9. Wang, Z., Rodrigues, M., Theilliol, D., Shen, Y.: Fault estimation filter design for discrete-time descriptor systems. IET Control Theory Appl. **9**(10), 1587–1594 (2015). https://doi.org/10.1049/iet-cta.2014.0641
10. Wang, Z., Shen, Y., Zhang, X., Wang, Q.: Observer design for discrete-time descriptor systems. An LMI approach. Syst. Control Lett. **61**(6), 683–687 (2012). https://doi.org/10.1016/j.sysconle.2012.03.006
11. Wu, Z.G., Su, H., Shi, P., Chu, J.: Analysis and Synthesis of Singular Systems with Time-Delays. Springer, Berlin (2013). https://doi.org/10.1007/978-3-642-37497-5
12. Xu, S., Lam, J.: Robust Control and Filtering of Singular Systems. Springer, Berlin (2006). https://doi.org/10.1007/11375753

Descriptor Principle in Residual Filter Design for Strictly Metzler Linear Systems

Dušan Krokavec and Anna Filasová

Abstract In the paper a descriptor technique to obtain residual filters for linear continuous-time Metzler systems is proposed. Conditions are introduced that achieve in design given levels of performance and structural compatibility with respect to existing parametric constraints. Resulting associated structure of linear matrix inequalities is outlined to possess stable asymptotic dynamic properties of the residual filters. The proposed design conditions are illustrated in the numerical example.

Keywords Linear Metzler systems · Fault detection · Descriptor-based approach · Residual filters · Linear matrix inequalities

1 Introduction

Positive system models are used in description of industrial variables pointing out to strictly positive quantities [4]. Restricting to Metzler structure of system matrices, these systems are often denoted as Metzlerian systems [13, 15]. Unfortunately known linear techniques cannot be straightly nominated to positive systems and so stable realisations of Metzler systems cover used specific methods [8, 16]. Other ways represent various methods based on linear programming, or a combination of linear programming with linear matrix inequalities [1]. Most recently, a principle based purely on the LMI formulation was presented in [10].

Fault detection filters, usually relying on the use of particular type of state observers, are mostly used to produce fault residuals in FTC. Because it is generally not possible in residuals to decouple totally fault effects from the perturbation influence, different approaches are used to tackle in part this conflict and to create

D. Krokavec (✉) · A. Filasová
Faculty of Electrical Engineering and Informatics, Department of Cybernetics and Artificial Intelligence, Technical University of Košice, Letná, 04200 Košice, Slovakia
e-mail: dusan.krokavec@tuke.sk

A. Filasová
e-mail: anna.filasova@tuke.sk

© The Author(s), under exclusive license to Springer Nature Switzerland AG 2021
J. Korbicz et al. (eds.), *Advances in Diagnostics of Processes and Systems*,
Studies in Systems, Decision and Control 313,
https://doi.org/10.1007/978-3-030-58964-6_6

residuals that are as a rule zero in the fault free case, maximally sensitive to faults, as well as robust to disturbances [3, 6].

Although the state observers received considerable attention, the descriptor design principle have not been studied extensively for non-singular systems [5]. The main issue of this paper is an application of the descriptor principle in residual filter design for strictly Metzler linear systems. Preferring LMI formulation, proofs use standard arguments in the sense of Lyapunov principle within set of LMIs representing structural constraints. Design conditions are formulated using strictly sharp LMIs, respecting diagonal stabilization principle for Metzler systems. Despite a partly conservative form, the design conditions and constraints are transformed to LMIs with minimal number of LMI variables.

The paper is organized as follows. Situated after Introduction, Sect. 2 gives a short description of main features of linear Metzler systems and Sect. 3 details the descriptor principle in Metzler state observer design within parametric performance requirements. LMI-based conditions ensuring the feasibility of residual filter design for given class of systems are derived in Sect. 4 and an illustrative example is provided to demonstrate the proposed approach in Sect. 5. Finally, Sect. 6 presents some concluding remarks.

Used notations are conventional so that x^T, X^T denotes transpose of the vector x and matrix X, respectively, $X \succ 0$ means that X is a symmetric positive definite matrix, the symbol I_n represents the n-th order unit matrix, $\rho(X)$ indicates the eigenvalue spectrum of a square matrix X, diag[·] enters up a block diagonal matrix, $\mathbb{R}$ ($\mathbb{R}_+$) qualifies the set of (nonnegative) real numbers, $\mathbb{R}^{n \times r}$ ($\mathbb{R}_+^{n \times r}$) refers to the set of $n \times r$ (nonnegative) real matrices and $\mathbb{M}_{-+}^{n \times n}$ means the set of Metzler square matrices.

2 Main Features of Linear Metzler Systems

To explain some properties, through this section there are considered strictly linear continuous-time Metzler systems given in the state-space form as

$$\dot{q}(t) = Aq(t) + Bu(t) + Ff(t), \tag{1}$$

$$y(t) = Cq(t), \tag{2}$$

where $q(t) \in \mathbb{R}_+^n$, $u(t) \in \mathbb{R}^r$, $y(t) \in \mathbb{R}_+^m$ are vectors of the system, input and output variables, respectively, $f(t) \in \mathbb{R}^p$ is the fault vector, and $A \in \mathbb{R}_{-+}^{n \times n}$, $B \in \mathbb{R}_+^{n \times r}$, $C \in \mathbb{R}_+^{m \times n}$, $F \in \mathbb{R}_+^{n \times p}$.

Non-negativity of matrices $B \in \mathbb{R}_+^{n \times r}$, $C \in \mathbb{R}_+^{m \times n}$, $F \in \mathbb{R}_+^{n \times p}$ implies from the Metzler system structure and means that all entries of these matrices are nonnegative and at least one is positive. The square Metzler matrix A is signum indefinite, but formally assignment $A \in \mathbb{R}_{-+}^{n \times n}$ highlights a Metzler matrix A.

Definition 1 ([2]) A square matrix $A \in \mathbb{R}^{n \times n}_{-+}$ is called strictly Metzler matrix if all its diagonal elements are negative and all its off-diagonal elements are positive. A Metzler matrix is stable if it is Hurwitz.

Systems with strictly Metzler structures is so confronted with n^2 boundaries implying from the Metzler matrix structural parametric constraints

$$a_{ii} < 0 \; \forall \, i = 1, 2, \ldots, n, \quad a_{ij,i \neq j} > 0 \; \forall \, i, j = 1, 2, \ldots, n . \tag{3}$$

It means that Metzler systems are diagonally stabilizable [14], while (3) can be implemented through linear matrix inequalities [10]. If non-diagonal elements of a Metzler matrix A are non-negative, it characterize non-strictly Metzler structure. Synthesis in this case requires consideration of further constraints [9, 11].

Proposition 1 ([4]) *A solution $q(t)$ of faulty-free (1) for $t \geq 0$ is positive and asymptotically stable if for non-negative $B \in \mathbb{R}^{n \times r}_+$ and strictly Metzler and Hurwitz $A \in \mathbb{R}^{n \times n}_{-+}$ the variable $q(t) \in \mathbb{R}^n_+$ when applying $u(t) \in \mathbb{R}^r_+$ and $q(0) \in \mathbb{R}_+$. The faulty-free system (1), (2) is asymptotically stable and positive if A is strictly Metzler and Hurwitz, $B \in \mathbb{R}^{n \times r}_+$, $C \in \mathbb{R}^{m \times n}_+$ are non-negative matrices and $y(t) \in \mathbb{R}^m_+$ for $u(t) \in \mathbb{R}^r_+$ and $q(0) \in \mathbb{R}_+$.*

Lemma 1 ([10]) *Let $A \in \mathbb{R}^{n \times n}_{-+}$, $B \in \mathbb{R}^{n \times r}_+$, $C \in \mathbb{R}^n_+$ while A is strictly Metzler, then faulty-free system (1), (2) is asymptotically stable if and only if there exists a positive definite diagonal matrix $S \in \mathbb{R}^{n \times n}_+$ such that for $i = 1, 2, \ldots, n, k = 1, 2, \ldots, n - 1$, the following set of linear matrix inequalities is feasible*

$$S \succ 0, \tag{4}$$

$$SA(i, i)_{(1 \leftrightarrow n)/n} \prec 0, \tag{5}$$

$$SL^k A(i + k, i)_{(1 \leftrightarrow n)/n} L^{k\mathrm{T}} \succ 0, \tag{6}$$

$$SA + A^{\mathrm{T}}S \prec 0, \tag{7}$$

where

$$A(i + k, i)_{(1 \leftrightarrow n)/n} = diag \begin{bmatrix} a_{1+k,1} & \cdots & a_{n,n-k} \; a_{1,n-k+1} & \cdots & a_{kn} \end{bmatrix} \tag{8}$$

and

$$L = \begin{bmatrix} \mathbf{0}^{\mathrm{T}} & 1 \\ I_{n-1} & \mathbf{0} \end{bmatrix}, \quad L^{-1} = L^{\mathrm{T}}, \tag{9}$$

$L \in \mathbb{R}^{n \times n}_+$ *is the circulant permutation matrix.*

The set of LMIs (5), (6) reflects the parametric structural constraints (3). Since P as well as $L^h A(i + k, i)_{(1 \leftrightarrow n)/n} L^{h\mathrm{T}}$ for all $k = 0, 1, 2, \ldots, n - 1$ are diagonal matrices, (5), (6) are symmetric. The inequality (7) guaranties that A is Hurwitz.

Lemma 2 ([7]) *Defining permutation matrix* $L \in \mathbb{R}^{n \times n}$ *in the circulant form (9) and considering a square diagonal matrix of dimension* $n \times n$ *as* $Z \in \mathbb{R}^{n \times n}$, *where*

$$Z = diag\left[\, z_{11} \; z_{22} \; \cdots \; z_{nn} \,\right],\tag{10}$$

then

$$L^{\mathrm{T}} Z L = diag\left[\, z_{22} \; \cdots \; z_{nn} \; z_{11} \,\right].\tag{11}$$

When one wish to have results on observer based fault residual filters, the descriptor principle in Metzler observer design is expressed at first.

3 Descriptor Principle in Metzler State Observer Design

Since state observer to fault-free system (1), (2) is given as

$$\dot{q}_e(t) = A q_e(t) + B u(t) + J(y(t) - y_e(t)),\tag{12}$$

$$y_e(t) = C q_e(t),\tag{13}$$

using (1), (2), (12), (13), it yields

$$\dot{e}(t) = (A - JC)e(t) + F f(t),\tag{14}$$

$$(A - JC)e(t) + F f(t) - \dot{e}(t) = 0,\tag{15}$$

respectively, where

$$e(t) = q(t) - q_e(t).\tag{16}$$

To apply for Metzler observer system matrix structure it yields

$$A_e = A - JC = A - \sum_{h=1}^{m} j_h c_h^{\mathrm{T}} = A - \sum_{h=1}^{m} J_{dh} l l^{\mathrm{T}} C_{dh},\tag{17}$$

where

$$J = \left[\, j_1 \; \cdots \; j_m \,\right], \; J_{dh} = \mathrm{diag}\left[\, j_h \,\right],\tag{18}$$

$$C = \begin{bmatrix} c_1^{\mathrm{T}} \\ \vdots \\ c_m^{\mathrm{T}} \end{bmatrix}, \; C_{dh} = \mathrm{diag}\left[\, c_h^{\mathrm{T}} \,\right], \; A = \begin{bmatrix} a_{11} & \cdots & a_{1n} \\ & \vdots & \\ a_{n1} & \cdots & a_{nn} \end{bmatrix}, \; l = \begin{bmatrix} 1 \\ \vdots \\ 1 \end{bmatrix}.\tag{19}$$

Using the descriptor principle, the following lemma presents the Metzler observer design conditions in terms of LMIs for fault-free system (1), (2).

Lemma 3 *Metzler state observer (12), (13) is stable if for given positive scalar $\delta \in \mathbb{R}$ there exist positive definite diagonal matrices $P, S, R_h \in \mathbb{R}^{n \times n}$ such that for $k = 1, 2, \ldots, n-1$ and $h = 1, 2 \ldots, m,$*

$$P \succ 0, \quad S \succ 0, \quad R_h \succ 0, \tag{20}$$

$$SA(i, i)_{(1 \leftrightarrow n)} - \sum_{h=1}^{m} R_h C_{dh} \prec 0, \tag{21}$$

$$SL^k A(i + k, i)_{(1 \leftrightarrow n)/n} L^{kT} - \sum_{h=1}^{m} R_h L^k C_{dk} T^{kT} \succ 0, \tag{22}$$

$$\begin{bmatrix} SA + A^T S - \sum_{h=1}^{m} (R_h ll^T C_{dh} + C_{dh} ll^T R_h) & * \\ P - \delta S + \delta SA - \delta \sum_{h=1}^{m} R_h C_{dh} & -2\delta S \end{bmatrix} < 0. \tag{23}$$

When the above conditions hold, observer strictly positive gain matrix J is determined as

$$J_{dh} = S^{-1} R_h, \quad j_h = J_{dh} l, \quad J = \begin{bmatrix} j_1 & \cdots & j_m \end{bmatrix}. \tag{24}$$

Hereafter, $*$ denotes the symmetric item in a symmetric matrix.

Proof An equivalent form to (15) with $f(t) = 0$ is given using the equality $\dot{e}(t) = \dot{e}(t)$ as

$$\begin{bmatrix} I_n & 0 \\ 0 & 0 \end{bmatrix} \begin{bmatrix} \dot{e}(t) \\ \ddot{e}(t) \end{bmatrix} = \begin{bmatrix} \dot{e}(t) \\ 0 \end{bmatrix} = \begin{bmatrix} 0 & I_n \\ A_e & -I_n \end{bmatrix} \begin{bmatrix} e(t) \\ \dot{e}(t) \end{bmatrix}, \tag{25}$$

which formally leads to

$$E^\circ \dot{e}^\circ(t) = A_e^\circ e^\circ(t), \tag{26}$$

where

$$e^{\circ T}(t) = \begin{bmatrix} e^T(t) & \dot{e}^T(t) \end{bmatrix}, \quad E^\circ = E^{\circ T} = \begin{bmatrix} I_n & 0 \\ 0 & 0 \end{bmatrix}, \quad A_e^\circ = \begin{bmatrix} 0 & I_n \\ A_e & -I_n \end{bmatrix}. \tag{27}$$

Defining the Lyapunov function of the form

$$v(e^\circ(t)) = e^{\circ T}(t) E^{\circ T} P^\circ e^\circ(t) > 0, \quad E^{\circ T} P^\circ = P^{\circ T} E^\circ \succeq 0, \tag{28}$$

the evaluation of time derivative of (28) along the trajectories of the observer error in the absence of faults gives

$$\dot{v}(e^\circ(t)) = \dot{e}^{\circ T}(t) E^{\circ T} P^\circ e^\circ(t) + e^{\circ T}(t) P^{\circ T} E^\circ \dot{e}^\circ(t) < 0 \tag{29}$$

and, inserting (26), this in turn implies that

$$\dot{v}(e^\circ(t)) = e^{\circ T}(t)(P^{\circ T}A_e^\circ + A_e^{\circ T}P^\circ)e^\circ(t) < 0, \tag{30}$$

$$P^{\circ T}A_e^\circ + A_e^{\circ T}P^\circ \prec 0, \tag{31}$$

respectively. Next, considering the case construction of P° that

$$P^\circ = \begin{bmatrix} P & T \\ S & Q \end{bmatrix}, \tag{32}$$

it can be proceed for the matrix relation from (28)

$$\begin{bmatrix} I_n & 0 \\ 0 & 0 \end{bmatrix}\begin{bmatrix} P & T \\ S & Q \end{bmatrix} = \begin{bmatrix} P^T & S^T \\ T^T & Q^T \end{bmatrix}\begin{bmatrix} I_n & 0 \\ 0 & 0 \end{bmatrix} \geq 0 \Rightarrow \begin{bmatrix} P & T \\ 0 & 0 \end{bmatrix} = \begin{bmatrix} P^T & 0 \\ T^T & 0 \end{bmatrix} \geq 0 \tag{33}$$

and it is evident that (33) can be satisfied only if

$$P = P^T \succ 0, \quad T = 0. \tag{34}$$

After simple manipulation (31) can be transformed into the following form

$$\begin{bmatrix} A_e^T S + S^T A_e & * \\ P - Q + Q^T A_e & -Q - Q^T \end{bmatrix} \prec 0 \tag{35}$$

and, owing to emerged products $S^T A_e$, $Q^T A_e$ in (35), the restriction on the structure of Q can be enunciated as $Q = \delta S$, where $\delta > 0$, $\delta \in \mathbb{R}_+$. Therefore,

$$Q^T A_e = \delta S^T(A - JC), \quad S^T(A - JC) = S^T A - \sum_{h=1}^m S^T J_{dh} ll^T C_{dh} \tag{36}$$

and since P, S have to be positive definite diagonal matrices, then with

$$R_h = SJ_{dh}, \tag{37}$$

(35) implies (23).

Since the components on the main diagonal of A_e are

$$a_{ell} = a_{ll} - \sum_{h=1}^r j_{lh}c_{hl} < 0, \quad l = 1, 2, \ldots, n, \tag{38}$$

with relation to (5), (8), (9) and (17)–(19) it can write

$$A_e(i, i)_{(1 \leftrightarrow n)/n} - \sum_{h=1}^{r} J_{dh} C_{dh} \prec 0. \tag{39}$$

Thus, multiplying the left side by S and using (37) then (39) implies (21).

The element below the main diagonal of A_e, with notation respecting summation modulo $n + 1$ [10], can be taken formally as

$$a_{e,l+1,l} = a_{l+1,l} - \sum_{h=1}^{r} j_{l+1,h} c_{h,1} > 0, \tag{40}$$

$$A_e(i + 1, i)_{(1 \leftrightarrow n)/n} - \sum_{h=1}^{r} J_{dh1} C_{dh} > 0, \tag{41}$$

respectively, where

$$J_{dh1} = \text{diag} \left[j_{2h} \cdots j_{nh} \ j_{1h} \right] \tag{42}$$

and using (11) it yields

$$A_p(i, i + 1)_{(1 \leftrightarrow n)/n} - \sum_{h=1}^{r} L^{\mathrm{T}} J_{dh} L C_{dh} > 0. \tag{43}$$

Pre-multiplying the right side by L^{T} and post-multiplying the left side by SL then (43) implies

$$SLA_p(i, i + 1)_{(1 \leftrightarrow n)/n} L^{\mathrm{T}} - \sum_{h=1}^{r} S J_{dh} L C_{dh} L^{\mathrm{T}} > 0 \tag{44}$$

and, using notation (37), it follows that

$$SLA_p(i, i + 1)_{(1 \leftrightarrow n)/n} L^{\mathrm{T}} - \sum_{h=1}^{r} R_h L C_{dh} L^{\mathrm{T}} > 0. \tag{45}$$

Referring to (22) it provides that (45) satisfies (22) for $k = 1$.

Analogously, one can deduce (22) using elements on the rest generalized diagonals of A_e. This concludes the proof. ∎

4   Residual Filter Design

Because the structure of Metzler systems introduces n^2 algebraic constraints into the synthesis of controllers as well as of observers, residual signals are, in general, sufficiently sensitive to additive faults even if the residual generator is defined in the simplest form

$$r(t) = Ce(t).$$
(46)

The following theorem is the current case to this construction but there is no problem to modify it, if necessary, for $r(t) = HCe(t)$, where H is a non-negative matrix of appropriated dimension.

Theorem 1 *The residual filter built on Metzler state observer (12), (13) is stable if for given positive $\delta \in \mathbb{R}_+$ there exist positive definite diagonal matrices $P, S, R_h \in \mathbb{R}^{n \times n}$ and a positive scalar $\gamma \in \mathbb{R}_+$ such that for $k \doteq 1, 2, \ldots, n-1$ and $h = 1, 2 \ldots, m$,*

$$P \succ 0, \quad S \succ 0, \quad R_h \succ 0,$$
(47)

$$SA(i, i)_{(1 \leftrightarrow n)} - \sum_{h=1}^{m} R_h C_{dh} \prec 0,$$
(48)

$$SL^k A(i + k, i)_{(1 \leftrightarrow n)/n} L^{kT} - \sum_{h=1}^{m} R_h L^k C_{dk} T^{kT} \succ 0,$$
(49)

$$\begin{bmatrix} SA + A^T S - \sum_{h=1}^{m} (R_h ll^T C_{dh} + C_{dh} ll^T R_h) & * & * & * \\ P - \delta S + \delta SA - \delta \sum_{h=1}^{m} R_h C_{dh} & -2\delta S & * & * \\ F^T S & \delta F^T S & -\gamma I_p & * \\ C & 0 & 0 & -\gamma I_m \end{bmatrix} \prec 0.$$
(50)

When the above conditions hold the residual filter parameters are given by (24).

Proof The assumption $f(t) \neq 0$ implies the following equivalently writing conditions

$$E^\circ \dot{e}^\circ(t) = A_e^\circ e^\circ(t) + F^\circ f(t),$$
(51)

$$r(t) = C^\circ e^\circ(t),$$
(52)

where

$$F^\circ = \begin{bmatrix} 0 \\ F \end{bmatrix}, \quad C^\circ = \begin{bmatrix} C & 0 \end{bmatrix}$$
(53)

and the Lyapunov function is considered with a positive $\gamma \in \mathbb{R}_+$ as

$$v(e^\circ(t)) = e^{\circ\mathrm{T}}(t)E^{\circ\mathrm{T}}P^\circ e^\circ(t) + \gamma^{-1}\int_0^t (r^\mathrm{T}(v)r(v) - \gamma^2 f^\mathrm{T}(v)f(v))dv > 0,$$

(54)

with the matrix relations defined as in (28). It can be readily inferred from (54) that

$$\dot{v}(e^\circ(t)) = \dot{e}^{\circ\mathrm{T}}(t)E^{\circ\mathrm{T}}P^\circ e^\circ(t) + e^{\circ\mathrm{T}}(t)P^{\circ\mathrm{T}}E^\circ \dot{e}^\circ(t) +$$
$$+ \gamma^{-1}r^\mathrm{T}(t)r(t) - \gamma f^\mathrm{T}(t)f(t) \qquad (55)$$
$$< 0.$$

Then, using (51), (52) the time derivative of $v(e^\circ(t))$ can be taken reformulated as follows

$$\dot{v}(e^\circ(t)) = e^{\circ\mathrm{T}}(t)(P^{\circ\mathrm{T}}A_e^\circ + A_e^{\circ\mathrm{T}}P^\circ + \gamma^{-1}C^{\circ\mathrm{T}}C^\circ)e^\circ(t) +$$
$$+ f^\mathrm{T}(t)F^{\circ\mathrm{T}}P^\circ e^\circ(t) + e^{\circ\mathrm{T}}(t)P^{\circ\mathrm{T}}F^\circ f(t) - \gamma f^\mathrm{T}(t)f(t) \qquad (56)$$
$$< 0.$$

Hence, if condition (56) holds, then it follows that

$$\begin{bmatrix} P^{\circ\mathrm{T}}A_e^\circ + A_e^{\circ\mathrm{T}}P^\circ + \gamma^{-1}C^{\circ\mathrm{T}}C^\circ & P^{\circ\mathrm{T}}F^\circ \\ F^{\circ\mathrm{T}}P^\circ & -\gamma I_p \end{bmatrix} \prec 0 \qquad (57)$$

and using Schur complement then (57) is equivalent to

$$\begin{bmatrix} P^{\circ\mathrm{T}}A_e^\circ + A_e^{\circ\mathrm{T}}P^\circ & P^{\circ\mathrm{T}}F^\circ & C^{\circ\mathrm{T}} \\ F^{\circ\mathrm{T}}P^\circ & -\gamma I_p & 0 \\ C^\circ & 0 & -\gamma I_m \end{bmatrix} \prec 0, \qquad (58)$$

where, since S is a positive definite diagonal matrix, it can consider $Q = \delta S, \delta > 0,$ $\delta \in \mathbb{R}_+$ and

$$P^{\circ\mathrm{T}}F^\circ = \begin{bmatrix} P^\mathrm{T} & S^\mathrm{T} \\ T^\mathrm{T} & Q^\mathrm{T} \end{bmatrix}\begin{bmatrix} 0 \\ F \end{bmatrix} = \begin{bmatrix} SF \\ \delta SF \end{bmatrix}. \qquad (59)$$

Thus, with the generated structure (30) the stability condition (58) takes the form

$$\begin{bmatrix} A_e^T S + S^T A_e & * & * & * \\ P - \delta S + \delta S A_e & -2\delta S & * & * \\ F^\mathrm{T}S & \delta F^\mathrm{T}S & -\gamma I_p & * \\ C & 0 & 0 & -\gamma I_m \end{bmatrix} \prec 0. \qquad (60)$$

Implementing (36), (37), the inequality (60) implies (50) and, because the parameter constraint stay unchanged, (21), (22) forced (48), (49). This concludes the proof. ∎

Remark 1 Reflecting $r(t) = HCe(t)$ then, by replacing in (60) C by HC, it is easy to modify LMI (50), to declare an additive matrix variable H and, finally, to construct residual generator matrix gain if a solution of such modified set of LMIs is feasible.

5   Illustrative Example

Let us consider the system in (1), (2) with the following data [12]

$$A = \begin{bmatrix} -3.380 & 0.208 & 6.715 & 5.676 \\ 0.581 & -4.290 & 2.050 & 0.675 \\ 1.067 & 4.273 & -6.654 & 5.893 \\ 0.048 & 2.273 & 1.343 & -2.104 \end{bmatrix}, \quad B = F = \begin{bmatrix} 0.0400 & 0.0189 \\ 0.0568 & 0.0203 \\ 0.0114 & 0.0315 \\ 0.0114 & 0.0170 \end{bmatrix},$$

$$C = \begin{bmatrix} 0 & 1 & 0 & 0 \\ 0 & 0 & 0 & 1 \end{bmatrix}, \quad l^{\mathsf{T}} = \begin{bmatrix} 1 & 1 & 1 & 1 \end{bmatrix}, \quad L = \begin{bmatrix} \mathbf{0}^{\mathsf{T}} & 1 \\ I_3 & \mathbf{0} \end{bmatrix}.$$

From synthesis formal consequence the matrix auxiliary parameters imply

$$A(i, i)_{(1 \leftrightarrow 4)/4} = -\mathrm{diag}\begin{bmatrix} 3.3800 & 4.2900 & 6.6540 & 2.1040 \end{bmatrix},$$

$$A(i + 1, i)_{(1 \leftrightarrow 4)/4} = \mathrm{diag}\begin{bmatrix} 0.5810 & 4.2730 & 1.3430 & 5.6760 \end{bmatrix},$$

$$A(i + 2, i)_{(1 \leftrightarrow 4)/4} = \mathrm{diag}\begin{bmatrix} 1.0670 & 2.2730 & 6.7150 & 0.6750 \end{bmatrix},$$

$$A(i + 3, i)_{(1 \leftrightarrow 4)/4} = \mathrm{diag}\begin{bmatrix} 0.0480 & 0.2080 & 2.0500 & 5.8930 \end{bmatrix}$$

$$C_{d1} = \mathrm{diag}\begin{bmatrix} 0 & 1 & 0 & 0 \end{bmatrix}, \quad C_{d2} = \mathrm{diag}\begin{bmatrix} 0 & 0 & 0 & 1 \end{bmatrix}.$$

The matrix variables P, S, R_1, R_1 satisfying (47)–(50) with interactive setting $\delta = 0.1$ are as follows (solved by SeDuMi package)

$$P = \mathrm{diag}\begin{bmatrix} 0.2705 & 1.0801 & 1.0304 & 1.1362 \end{bmatrix},$$

$$S = \mathrm{diag}\begin{bmatrix} 0.9421 & 1.6165 & 1.5829 & 2.3327 \end{bmatrix},$$

$$R_1 = \begin{bmatrix} 0.0950 & 1.8443 & 5.0191 & 3.4822 \end{bmatrix},$$

$$R_2 = \begin{bmatrix} 4.4469 & 0.5811 & 7.4607 & 2.3893 \end{bmatrix}$$

and obviously reflect positive definite diagonal structures when solving for maximized H_∞ norm upper bound of residual filter transfer function $\gamma = 3.7143$.

Thus, with the matrices S, R_1, R_2 given as above, (24) implies the strictly positive residual filter gain

$$
J = \begin{bmatrix} 0.1008 & 4.7200 \\ 1.1410 & 0.3595 \\ 3.1709 & 4.7134 \\ 1.4928 & 1.0243 \end{bmatrix}
$$

and the filter system structure is with strictly Metzler and Hurwitz stable matrix

$$
A_e = \begin{bmatrix} -3.3800 & 0.1072 & 6.7150 & 0.9560 \\ 0.5810 & -5.4310 & 2.0500 & 0.3155 \\ 1.0670 & 1.1021 & -6.6540 & 1.1796 \\ 0.0480 & 0.7802 & 1.3430 & -3.1283 \end{bmatrix}, \quad \rho(A_e) = \begin{Bmatrix} -8.6251 \\ -5.4536 \\ -3.5576 \\ -0.9569 \end{Bmatrix}.
$$

Since the first and the third columns of C are zero ones, it can see that the first and the third columns of A and A_e are identical. Therefore, for this structure C, the principle can also be applied to a Metzler system that would have (some) non-diagonal elements equal to zero in these columns.

To apply in simulations, the system can be stabilized by state feedback control $u(t) = -Kq(t) + Ww(t)$, with the gain structures [12]

$$
K = \begin{bmatrix} 1.9171 & 1.3770 & 21.8659 & 1.1329 \\ 0.8458 & 5.2546 & 31.1965 & 27.8961 \end{bmatrix}, \quad W = \begin{bmatrix} 157.8203 & -91.7855 \\ -229.3141 & 157.6823 \end{bmatrix}
$$

and $w^T(t) = [0.5 \; 0.6]$, $q(0) = q_e(0) = 0$.

The basic results of system state estimation for the first fault acting on linear Metzlerian system are visualized in Fig. 1, where

$$
B_{f1} = \begin{bmatrix} 0.995b_1 & b_2 \end{bmatrix}, \quad B_{f2} = \begin{bmatrix} b_1 & 0.995b_2 \end{bmatrix}.
$$

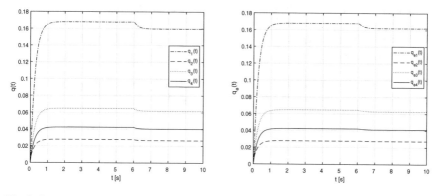

Fig. 1 System under the first actuator fault—state and state estimations

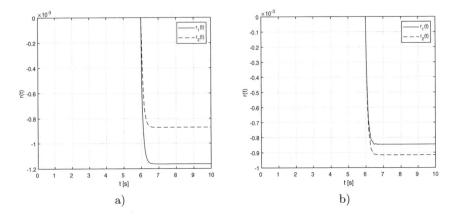

Fig. 2 Fault residual filter responses

It can see from the figure the perfect system state estimate also in considered faulty regime.

The left part (a) of Fig. 2 shows the evolution of the residual signals generated by fault residual filter as a response on the fault starting its step-like action on the first actuator at time instant $t = 6$ s. The right part (b) illustrates residual response to fault acting on the second actuator and is simulated in the example by the same scenario. The residual signal time responses document high sensitivity of residual signals to additive faults.

6   Concluding Remarks

Based on descriptor approach a new residual filter design method for continuous-time Metzlerian linear systems is introduced in the paper. Presented version is derived in terms of optimization over LMI defined parametric constraints by standard numerical procedures to manipulate the filter asymptotic stability, dynamics and positiveness. The numerical results show asymptotic stability for bounded faults and good structural performances.

Acknowledgements The work presented was supported by VEGA, the Grant Agency of the Ministry of Education and the Academy of Science of Slovak Republic under Grant No. 1/0608/17. This support is gratefully acknowledged.

References

1. Ait Rami, M., Tadeo, F.: Linear programming approach to impose positiveness in closed-loop and estimated states. In: Proceedings of the 16th International Symposium on Mathematical Theory of Networks and Systems, Kyoto, Japan, pp. 2470–2477 (2006)
2. Berman, A., Neumann, M., Stern, R.: Nonnegative Matrices in Dynamic Systems. Wiley, New York (1989). https://doi.org/10.1016/0378-4754(91)90046-6
3. Chen, W., Saif, M.: Observer-based strategies for actuator fault detection, isolation and estimation for certain class of uncertain nonlinear systems. IET Control Theory Appl. 1(6), 1672–1680 (2007). https://doi.org/10.1049/iet-cta:20060408
4. Farina, L., Rinaldi, S.: Positive Linear Systems. Theory and Applications. Wiley, New York (2000). https://doi.org/10.1002/9781118033029
5. Fridman, E., Shaked, U.: A descriptor system approach to H_∞ control of linear time-delay systems. IEEE Trans. Autom. Control 47(2), 253–270 (2002). https://doi.org/10.1109/9.983353
6. Guo, J., Huang, X., Cui, Y.: Design and analysis of robust fault detection filter using LMI tools. Comput. Math. Appl. 57(11–12), 1743–1747 (2009). https://doi.org/10.1016/j.camwa.2008.10.032
7. Horn, R.A., Johnson, C.R.: Matrix Analysis. Cambridge University Press, New York (1985). https://doi.org/10.1017/CBO9780511810817
8. Kaczorek, T.: Stability and stabilization of positive fractional linear systems by state feedback. Bull. Pol. Acad. Sci. Tech. Sci. 58(4), 537–554 (2010). https://doi.org/10.2478/v10175-010-0054-y
9. Krokavec, D., Filasová, A.: Stabilization of discrete-time LTI positive systems. Arch. Control Sci. 27(4), 575–594 (2017). https://doi.org/10.1515/acsc-2017-0034
10. Krokavec, D., Filasová, A.: LMI based principles in strictly Metzlerian systems control design. Math. Probl. Eng. 2018, 1–14 (2018). https://doi.org/10.1155/2018/9590253
11. Krokavec, D., Filasová, A.: H_∞ norm principle in residual filter design for discrete-time linear positive systems. Eur. J. Control 45, 17–29 (2019). https://doi.org/10.1016/j.ejcon.2018.10.001
12. Krokavec, D., Filasová, A.: Mixed H_2/H_∞ strategy in control law parameter design for linear strictly Metzlerian systems. In: Proceedings of the 8th International Conference on Systems and Control ICSC 2019, Marrakech, Morocco, pp. 476–481 (2019). https://doi.org/10.1109/ICSC47195.2019.8950497
13. Nikaido, H.: Convex Structures and Economic Theory. Academic Press, New York (1968)
14. Shorten, R., Mason, Q., King, C.: An alternative proof of the Barker, Berman, Plemmons (BBP) result on diagonal stability and extensions. Linear Algebra Appl. 430(1), 34–40 (2009). https://doi.org/10.1016/j.laa.2008.06.037
15. Smith, H.L.: Monotone Dynamical Systems. An Introduction to the Theory of Competitive and Cooperative Systems. American Mathematical Society, Providence (1995). https://doi.org/10.1090/surv/041
16. Son, N.K., Hinrichsen, D.: Robust stability of positive continuous time systems. Numer. Funct. Anal. Optim. 17(5–6), 649–659 (1996). https://doi.org/10.1080/01630569608816716

Hierarchical Model for Testing a Distributed Computer System

A. M. Gruzlikov and N. V. Kolesov

Abstract The approach to test-based diagnosis of distributed computer systems considered in the paper involves introduction of redundancy in the system aimed to make the test design and the diagnostic process simpler. In effect, the introduced redundancy is a diagnostic model of the system, for which sufficient conditions of controllability and observability are formulated. The diagnostic model runs in parallel with the main software of the system.

Keywords Diagnosis · Hierarchical model · Parallel computations · Real time systems · Non-stationary models

1 Introduction

Problems of diagnosis and fault tolerance are of high priority for designers of control systems and information processing. Despite the fact that research in this area has been carried out for several decades and the results are reported in numerous publications, practice brings new challenges that require in-depth study. Among the most significant factors stimulating the new research are the continuously increasing complexity of systems and multiplicity of causes of malfunctioning, including hardware failures and errors in software. In practice, the difficulties associated with high dimensionality of the problem are overcome by its decomposition on the basis of a hierarchical approach, wherein the system components are distributed in accordance with the levels of complexity. Further, specific diagnostic tools (DT) are developed for each level to detect faults. Practical solutions used to synthesize DT rely on the techniques of functional and test diagnosis [8, 9, 11, 12].

The mathematical model of the diagnosed object makes use of the Petri net, dynamical system and the finite-state machine [1, 2, 13]. It is known that solving a

A. M. Gruzlikov (✉) · N. V. Kolesov
State Research Center of the Russian Federation – Concern CSRI Elektropribor JSC,
30, Malaya Posadskaya Street, Saint Petersburg 197046, Russia
e-mail: agruzlikov@yandex.ru

© The Author(s), under exclusive license to Springer Nature Switzerland AG 2021
J. Korbicz et al. (eds.), *Advances in Diagnostics of Processes and Systems*,
Studies in Systems, Decision and Control 313,
https://doi.org/10.1007/978-3-030-58964-6_7

Fig. 1 A data flow graph of
a real-time system S (**a**),
structure of the redundant
system S with diagnostic
tool (**b**)

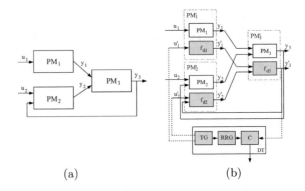

(a)   (b)

diagnostic problem requires certain redundancy in hardware (software), information, or time. DTs are an example of such redundancy. In addition, prior to the synthesis of DTs, redundancy can be introduced into the system to make the diagnosis process simpler: increase in the number of the system inputs or outputs, introduction of special tools for the system analysis, etc. The introduced redundancy can formally relate both to the system and DTs, which is insignificant. It is to this line of research—introduction of redundancy with the aim to make the diagnosis easier—that the approach proposed earlier in [4–7] and developed by the authors in this work belongs.

In this paper we consider distributed real-time computing systems, which are the core of modern control and data-processing systems, a characteristic feature of which is processing information is asynchronously and a periodic input data flow. The discussion focuses on the formation of top-level diagnostic model, wherein the distributed system is represented as a composition of program modules (PM) that make up the functional software, and the failure class considered includes all kinds of faults of addressing between the PMs.

The diagnostic model under study is focused on test diagnosis; actually, it is a linear finite-state machine (binary linear dynamic system). A feature of the approach under consideration is that the diagnostic model is integrated into the system; it is executed in parallel with the main software of the system and makes the process of its test diagnosis simpler. Taking the foregoing into account, the approach was called diagnostics with a parallel model. From the class-of-failures standpoint, the proposed diagnostic model can be associated to discrete event systems, because the system operation is represented in the language of sequences events. In this case, its are events of information exchange between PMs. Similar models are widely used in the analysis and testing of complex systems [12].

The main contribution of this study is formulation of sufficient conditions for observability and controllability for advanced models of a more general form as compared with those described in previous publications [5, 7]. Using advanced models reduces the amount of diagnostic information transmitted between PMs. The properties of the model observability and controllability significantly affect the complexity of the procedures for constructing tests.

2 Preliminary Information

Let us describe briefly the approach to testing of DS with a parallel model [4, 5] and illustrate it with a simple example. Figure 1a shows a data flow graph of a real-time system S. The system has three functionally related program modules: PM_1, PM_2, and PM_3, which can be placed both on different processors of the system, and on a single one. Using the input data, each of the PMs (u_1—for PM_1, u_2 and y_3—for PM_2, and y_1 and y_2—for PM_3) forms the output (y_1, y_2 and y_3, respectively).

Input data arrive and are processed in real time with a certain specified period. In the system, all input flows of a particular PM are the arguments of the function it implements, which are needed for its calculation. In Fig. 1b, system S is shown together with DT. The system has the redundancy introduced in it, which is represented by algorithms f_{d1}, f_{d2} and f_{d3}. DTs consist of a test generator (TG), a reference-response generator (RRG), and a comparator (C). DTs form test data for the system, complementing them with input data for PM_1 and PM_2, and exam the output test data of system S (PM_3). Through the exchange channels, each PM_1, PM_2 and PM_3 receives information which is processed by the main algorithms. In parallel with this, test information words are processed by algorithms f_{d1}, f_{d2} and f_{d3} that react to the events of information transmit and the results of their processing are included in the output data (y_3). Since the mechanism for exchanging real and test data in the system is common, it becomes possible, based on the test results obtained during operation, to conclude on the presence or absence of faults in the addressing of data transmit. Note that distortions of real data during the exchange process are not included in the class of faults considered here. Thus, in the case of testing with a parallel model, the problem consists in designing algorithms for test processing in the PM, as well as designing the tests themselves in accordance with the content of the algorithms.

In the authors' opinion, the relevance of this approach is dictated by the following simple considerations. First of all, it should be noted that the DT at the upper level could be realized without using a parallel model, but only relying on the main software. Obviously, such a solution will be much more complicated both at the stage of the test design and at the stage of its practice. When designing a test, the developer will have to work with complicated, usually non-linear, software. It is clear that this procedure will be characterized by an exponential dependence on the system dimension, since such complexity is characteristic of similar procedures for finite-state machines of a general form [10]. In addition, this step will require allocation of a time interval for the test on the time diagram in contrast to the case with a parallel model. In the case of a parallel model, the test construction and it processing are based on a linear model characterized by a polynomial dependence on the dimension of the system. The relative simplicity of obtaining observable and controllable models in the linear case has a profound effect on reducing the complexity of the test design, which is extremely important for the construction of effective tests. Note that for the main non-linear software, these properties are usually rather limited and difficult to analyze.

The basis of the approach to testing DS with a parallel model is an algorithm for synthesizing a diagnostic model with a hierarchical structure that consists of two stages. At the first stage of the model construction, well-known algorithms are used to form a set of computational paths [3]. These paths make up the coverage of the graph directed links of data flow. In this context, a computational path is used in reference to a sequence of triggered PMs that connects a certain input with an output. Then, a chain, including as many dynamic links as there are PMs through which the path passes on, is assigned to each path. After the described procedures, the system model is represented as a set of functionally independent chains, and the diagnostic problem can be reduced to the diagnosis of individual chains.

It is taken into account that the sought dynamic model of the system is further used to design tests and that the procedure for constructing tests is simplified if the system model is, firstly, linear, and secondly, controllable and observable. Hence, we can formulate the requirement for the links of the model chains: they must be linear, that is,

$$x_i(t+1) = f_i x_i(t) + g_i u_i(t), \quad y_i(t) = h_i x_i(t) \quad i = 1 : v \tag{1}$$

where $x_i(t)$, $u_i(t)$, $y_i(t)$ are vectors of state, input and output, f_i, g_i, h_i are matrices of dynamics, input and output, correspondingly, for the i-th link of the chain, and v is the number of links in the model. In addition, the links must be such that the system model could also be controllable and observable.

The structure of the model with independent chains for the example under consideration is shown in Fig. 2a. However, in some cases, the application of such a DS model may require transfer of a large amount of diagnostic information through the exchange channels, which is not always permissible. In such situations, it is advisable to process several arrays of information by one link (merging of computational paths) [7].

This option is illustrated in Fig. 2b. Here, the output arrays of information of links M_{11} and M_{22} are processed by link M_{31}, while link M_{32} is excluded. As a result, the dimensionality of the output vector is reduced, and, consequently, the amount of diagnostic information from the PM in each tack of the input data processing is also reduced.

A dynamic description of a chain or a model with chain merging is obtained in accordance with the following rules. Assume that only one exchange takes place in a system at each instant of time. Then, state vector $x(t)$ is formed, which is composed

Fig. 2 A model structure for a system with independent chains (**a**), with merging of chains (**b**)

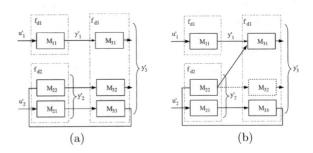

(a)   (b)

of the state vectors of links $x_i(t)$ $i = 1 : v$ included in this model. Matrices $F(t)$, $G(t)$, $H(t)$ are used to describe the transfer of information between the PM and the DT in each j-th information exchange. This will happen if the assumption of the uniqueness of the exchange at each instant of time is fulfilled.

In practice, this assumption is not fulfilled; however, as shown in [7], this is not an obstacle to using such models in designing tests. For convenience of description, each sequence of matrices on an interval equal to a period is related to its own sequence of indices, the set of which is denoted as $\Gamma = \{\gamma_s | s = 1 : N\}$, where N is the total number of exchanges in the system. Their initial segments on an interval with a period duration are obtained as a result of a cyclic shift of the sequence of indices $1, 2, \ldots, N$. At $N = 3$, there are three sequences of indices: $\Gamma = \{(1, 2, 3); (2, 3, 1); (3, 1, 2)\}$. Then

$$x(t + 1) = F(\gamma_s(j))x(t) + G(\gamma_s(j))u(t), \quad y(t) = H(\gamma_s(j))x(t) \quad j = 1 : N \tag{2}$$

The matrices in these equations depend on the identifier of the data exchange between the PMs, i.e., the model is non-stationary. Moreover, it is periodically non-stationary since the processes of data processing in the system are periodic because of the periodicity of the input flow. Recall that the properties of observability and controllability for non-stationary systems depend on the intervals γ_s used for observation and control, which is why it is appropriate to speak of γ_s-observability and γ_s-controllability.

Example 1 The formation of the model dynamics can be illustrated by the example of a merging of two chains (Fig. 2b). Let us introduce the following numeration for chain links, and, consequently, for the matrices and vectors describing them: $1 - M_{11}, 2 - M_{21}, 3 - M_{33}, 4 - M_{22}, 0 - M_{31}$. Assuming the following format of the model state vector $x = [x_1 \, x_2 \, x_3 \, x_4 \, x_0]^T$, we have the following sequence of exchange matrices on a period that ends with a session of test information reception from the DT (zero matrix elements are not shown):

$$F_{1,0}(1) = \begin{bmatrix} I & & & & \\ & I & & & \\ & & I & & \\ & & & I & \\ g_0h_1 & & & & f_0 \end{bmatrix}, \quad F_{2,3}(2) = \begin{bmatrix} I & & & & \\ & I & & & \\ & g_3h_2 & f_2 & & \\ & & & I & \\ & & & & I \end{bmatrix},$$

$$F_{3,4}(3) = \begin{bmatrix} I & & & & \\ & I & & & \\ & & I & & \\ & & g_4h_3 & f_4 & \\ & & & & I \end{bmatrix}, \quad F_{4,0}(4) = \begin{bmatrix} I & & & & \\ & I & & & \\ & & I & & \\ & & & I & \\ & & & g_0h_4 & f_0 \end{bmatrix}, \quad F_{0,DT}(5) = I$$

Recall some properties of the model under study or the features of its application formulated and used in previous works [7] since they will be helpful in further discussion.

Property 1 A difference is allowed in the temporal positioning of internal and, as a result, output events of information processing in this model and in the real system. This difference is due to the lack of information on the actual duration of PM execution duration during data processing, which it does not prevent from comparing these events and diagnosing. Owing to this feature, there is have a certain freedom in choosing a model. In particular, during the formation of a chain model, it allows us to assume that different pieces of information are processed not in parallel, but strictly sequentially. In the model with chain merging, it is assumed that not only different pieces of information are processed sequentially, but also the same piece of information in different chains that underwent merging. Following the described order, the state vectors of the chains are also ordered in the state vector of structure $\hat{S}^f$. In this case, the state vector of the merging link is the last in the queue. The above considerations make it possible to describe a parallel computing process in a distributed system by means of a sequential process, so that only one session is executed at a time and described by the dynamic system.

Property 2 Any periodically non-stationary model from the class under consideration can be reduced to a stationary one. This possibility is, first of all, due to the fact that both in the case of a model with independent chains and a model with chain merging, it is possible to change the DS description by the asynchronous model for a description by the synchronous sequential model. In this case, the non-stationary system with independent chains changes into a stationary system with independent chains, and the non-stationary system with chain merging, into a stationary system with chain merging, namely: for any synchronous periodically non-stationary system S^m described by the model

$$x(t+1) = F(\gamma_s(j))x(t) + G(\gamma_s(j))u(t), \ y(y) = H(\gamma_s(j))x(t), \ j = 1:N, \ (3)$$

the one that has, over the period of its performance, one session of information reception from the DT described by matrices $F(\gamma_r(N)), G(\gamma_r(N)), H(\gamma_r(N))$ for a certain sequence and one session of information output to the DT described by matrices $F(\gamma_r(N-1)), G(\gamma_r(N-1)), H(\gamma_r(N-1))$, there is a synchronous stationary system $\hat{S}^m$

$$x(k+1) = Ax(k) + Bu(k), \ y(k) = Cx(k) \qquad (4)$$

which, at any input sequence, forms output sequences that coincide with the output sequences of system S^m in γ_r. In this case, k are the periods of system (3), $A = F(\gamma_r) = \prod_{i=1}^{N} F(\gamma_r(N-i+1))$, $B = G(\gamma_r(N))$, $C = H(\gamma_r(N-1))F^{-1}(\gamma_r(N))A = H(\gamma_r(N-1))A$.

3   Problem Statement

Below, we consider the following problem. It is assumed that system S to be diag-nosed has a hierarchical structure of the model with an arbitrary number of merging points. This structure can be obtained from a model with independent chains by introducing a certain set of merging points. It is necessary to formulate sufficient conditions which the links of the model must satisfy and under which the system model will be observable and controllable.

The search for conditions is carried out under the following assumptions:

1. Models S^m and $\hat{S}^m$ are described in a binary field (binary arithmetic).
2. System S^m includes only links with scalar inputs and outputs, which reduces the amount of transmitted diagnostic information, but, at the same time, makes the problem of ensuring observability and controllability of the system maximally complicated as a whole.
3. To simplify the analysis, it is assumed that in the composite state vector of model $\hat{S}^m$, the state vectors of the links are numbered in the direction from the input to the output of the chain.

The stationary models $\hat{S}^m$ considered below can contain structural elements of three types: chain $\hat{S}^c$ (Fig. 3a), structure $\hat{S}^f$ with merging chains $\hat{S}^f$ (Fig. 3b), and hierarchical structure $\hat{S}^h$ (Fig. 3c), in which each of subsystems S_i can also be either a chain or a structure with chain merging, or hierarchical structure.

4   The Conditions of Observability and Controllability of a Parallel Model with More Than One Merging Point in It

Let us formulate the conditions sufficient for observability and controllability of the diagnostic model under consideration in the case that there is more than one merging point. In doing so, we follow the logic of the proof given in [7], which consists in the following. First, let us take the advantage of changing a periodically non-stationary

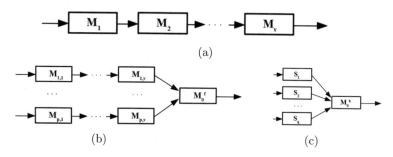

Fig. 3   Types of structural elements

model for a stationary one (Property 1). Then, we prove sufficient conditions that the links of the original non-stationary model must satisfy in order for the stationary model to be observable and controllable. After that, we show that these conditions are also sufficient for the observability and controllability of the non-stationary model.

Theorem 1 *The dynamics matrix A^c of the stationary chain $\hat{S}^c$ (Fig. 3a) of v links (1) is a non-singular lower block-triangular matrix with a characteristic polynomial $\phi^c = \prod_{i=1}^{v} \phi_i$, where ϕ_i is the characteristic polynomial of the i-th link $i = 1 : v$, which has the form:*

$$
A^c = \begin{bmatrix}
f_1 & & & & \\
R_{1,2} & f_2 & & & \\
& & \ddots & & \\
& & & f_{v-1} & \\
& & & R_{v-1,v} & f_v
\end{bmatrix},
$$

where f_i is the dynamics matrix of the i-th link $i = 1 : v$; $R_{i,i+1} = g_{i+1}h$, is the matrix of connection between the i-th and $(i + 1)$-th links $i = 1 : v - 1$.

Proof By the definition and in accordance with the second property, the session matrix in the non stationary model is a lower block-triangular matrix with blocks f_i $i = 1 : v$ and $R_{i,j+1}$ $i = 1 : v - 1$. Each of its blocks on the main diagonal, except one, contains unit matrices I.

The only "nonsingular" block that describes the exchange between the links M_i and M_{i+1}, exchanging information in this session, has the form:

$$
F_{i,i+1} = \begin{bmatrix}
I & 0 \\
g_{i,i+1} & f_{i+1}
\end{bmatrix},
$$

where $g_{i,i+1}$ is the matrix of connection between the i-th and $(i + 1)$-th links.

As a result, the dynamics matrix A of the stationary model is also a lower block-triangular matrix, since it is formed as a product of the session matrices. Since each link is triggered only once over a period during the reception of information, the main diagonal of matrix A will have the dynamics matrix of the chain links in accordance with the link order in the chain. It is known that the characteristic polynomial of the lower block-triangular matrix is equal to the product of the characteristic polynomials of the blocks on the main diagonal. Since the link dynamics matrices are nonsingular, matrix A^c will also be nonsingular.

Theorem 2 *The dynamics matrix A^f of the stationary structure $\hat{S}^f$ with merging of p chains (Fig. 3b) is a non-singular lower block-triangular matrix with a characteristic polynomial l $\phi^f = \phi_0^p \prod_{j=1}^{p} \phi_j^c$, where ϕ_j^c is the characteristic polynomial of the merging link M_0^f to power p; ϕ_j^c is the characteristic polynomial of the j-th chain $j = 1 : p$, which has the form:*

$$
A^f = \begin{bmatrix}
A^c_{\gamma_r,1} & & & & \\
& A^c_{\gamma_r,2} & & & \\
& & \cdots & & \\
& & & A^c_{\gamma_r,p} & \\
R_{1,0} & R_{2,0} & \cdots & R_{p,0} & f^p_0
\end{bmatrix},
$$

where $A^c_{\gamma_r,l}$ is the dynamics matrix of the l-th chain; $R_{l,0}$ is the matrix of connection between the l-th chain $l = 1 : p$ and the merging link.

Proof The formation of chain dynamics matrices was considered in the discussion of the previous proposition. They are placed along the main diagonal A^f in accordance with the specified order. Since all matrices $A_{\gamma_r,l}, l = 1 : p$ as well as matrix f^p_0 are nonsingular, matrix A^f is also nonsingular. Power p of the matrix f_0 occurrence in the expression for A^f is explained by the fact that the merging link is triggered p times over a period (receiving information p times).

Theorem 3 *The dynamics matrix A^h of the stationary hierarchical structure $\hat{S}^h$ (Fig. 3c) is a non-singular lower block-triangular matrix with a characteristic polynomial $\phi^h = \phi^q_0 \prod^q_{k=1} \phi^s_k$, where ϕ^q_0 is a characteristic polynomial of the merging link M^h_0 to power q; ϕ^s_k is a characteristic polynomial of subsystem S_k $k = 1 : q$. Matrix A^h has the form:*

$$
A^h = \begin{bmatrix}
A^s_{\gamma_r,1} & & & & \\
& A^s_{\gamma_r,2} & & & \\
& & \cdots & & \\
& & & A^s_{\gamma_r,p} & \\
R_{1,0} & R_{2,0} & \cdots & R_{q,0} & f^q_0
\end{bmatrix},
$$

where the dynamics matrix $A^s_{\gamma_r,l}$ corresponds to the l-th subsystem.

Proof It is easy to see that the proof of the previous proposition can be used as the proof for this one as well. Indeed, any subsystem can be treated as certain link. Thus, in this case we deal with the situation with merging of single-link chains that has been considered before.

So, the above theorems formulate simple rules for calculating characteristic polynomials of chains based on characteristic polynomials of links, characteristic polynomials of structures with merging chains based on characteristic polynomials of chains and links, and finally, characteristic polynomials of hierarchical structures based on characteristic polynomials of all types of structures. It is important because the conditions of observability and controllability of the models under consideration are formulated in terms of their characteristic polynomials. For chains and models with chain merging, these conditions are given in [7]; therefore, we derive them for a hierarchical structure of an arbitrary form.

Theorem 4 *The hierarchical stationary structure $\hat{S}^h$ consisting of observable and controllable scalar subsystems $\{S_k | k = 1 : q\}$ is observable and controllable if all* *polynomials from the following set are coprime: characteristic polynomials $\{\phi_k | k = 1 : q\}$ of subsystems, characteristic polynomial ϕ_0 and polynomials $\{c_k | k = 1 : q\}$ of the numerators of transfer functions $\{J_k | k = 1 : q\}$ of the merging link M_0^k.*

Proof First, let us show that the hierarchical structure $\hat{S}^h$ is observable. Since the merging link and all subsystems are observable, we can use make use of their description in a row observable form. It is known [7] that this description can be the foundation stone for obtaining a basis of space for periodic output sequences of free motion for any system. This basis is given on a interval of time equal to a period and is called reproducing matrix V, the size of which is $n \times T$ (n is the system dimension, T is a period of the output sequences of the system free motion):

$$
V = \begin{bmatrix}
\alpha_0 \; \alpha_1 \; \ldots \; \alpha_{T-n-1} & \alpha_{T-n} & & \\
& \alpha_0 \; \alpha_1 & \ldots & \alpha_{T-n-1} \; \alpha_{T-n} \\
& & \ldots & \\
& \alpha_0 & \alpha_1 & \ldots \; \alpha_{T-n-1} \; \alpha_{T-n}
\end{bmatrix},
$$

Here, $\{\alpha_i | i = 0 : T - n\}$ are the coefficients of the reproducing polynomial $\Psi(x)$, which is determined by the formula $\Psi(x) = \frac{1+x^T}{\phi(x)}$; $\phi(x)$ is the characteristic polynomial of the system dynamics matrix.

Let us show that in the case of free motion, the basis of the space of the output sequences of the hierarchical structure $\hat{S}^h$ given on the period T (T is a period of the output sequences of free motion of the hierarchical structure $\hat{S}^h$) consists, firstly, of the basis of the subspace of the output sequences of the merging link free motion and, secondly, of the bases of similar subspaces of the subsystems to be combined, processed by the merging link and given on the same period.

The first basis—the basis of subspace P_0 of the output sequences of the merging link M_0^h free motion—is represented by the reproducing matrix V_0 of this link repeated $\frac{T}{T_0}$ times on the period T, where T_0 is a period of the output sequence of the merging link free motion, i.e., the basis has the form $[V_0 V_0 \ldots V_0]$. The polynomial describing the first vector from this basis is calculated by the formula:

$$
\Psi(x) = \frac{1 + x^{T_0}}{\phi_0(x)} (1 + x^{T_0} + x^{2T_0} + \cdots + x^{T-T_0}) = \frac{1 + x^T}{\phi_0(x)} \tag{5}
$$

The polynomial for the second vector is obtained from (5) by multiplying by x, for the third vector—by x^2, etc., and for the last one—by x^{m_0-1} (m_0 is the dimension of the merging link).

Further, to describe the remaining bases—the bases of the output-sequence subspace of the subsystems free motion processed by the merging link—it is convenient to use the well-known d-transformation [7] for the sequences over Galois fields. In this case, the sequences are represented in the form of power polynomials, and for dynamic systems, we introduce the notion of a transfer function. Let us write the expression for the transfer function for the linear dynamic merging link (6). This function will be generally unique for each of the subsystems to be combined, because for different subsystems, the input matrices of the connecting link are different. Taking into account the rules of calculations in a binary field and introduction of variable x instead of d, we can write this function as:

$$J_k = h_0(xf_0 + I)^{-1}xg_k = \frac{h_0(xf_0 + I)^P xg_k}{|xf_0 + I|} \tag{6}$$

where $(xf_0 + I)^P$ is the matrix of algebraic complements expressed by power polynomials; $|xf_0 + I|$ is the polynomial $\phi^*_{f_0}(x)$ dual to the characteristic polynomial $\phi_{f_0}(x)$.

The basis of subspace P_k of the output sequences of the k-th subsystem free motion is represented by the reproducing matrix V_k of this subsystem repeated $\frac{T}{T_k}$ times on the period T, where T_k is a period of the output sequence of the k-th subsystem free motion, i.e., the basis has the form $[V_k V_k \ldots V_k]$.

Following (6), but taking into consideration the processing in the merging link, we obtain a polynomial describing the first vector from the basis of the k-th subsystem:

$$\Psi^{''}_k(x) = \frac{1 + x^{T_k}}{\phi_k(x)}(1 + x^{T_k} + x^{2T_k} + \cdots + x^{T-T_k})J_k = \frac{1 + x^T}{\phi_k(x)}J_k. \tag{7}$$

Substituting $J_k = \frac{c(x)}{\phi^*_k(x)}$ into (7), we derive:

$$\Psi^{''}_k(x) = \frac{1 + x^{T_k}}{\phi_k(x)}\frac{c(x)}{\phi^*_k}. \tag{8}$$

Let us show that the basis vectors from subspaces P_0 and P_k $k = 1 : q$ are linearly independent. Assuming the existence of a linear relationship, we derive the expression:

$$a(x)\Psi^{'}_{f^p_0}(x) + \sum_{k=1}^{q} b_k(x)\Psi^{''}_k(x)J_k(x) = 0, \tag{9}$$

where $a(x)$ and $b_k(x)$ are polynomials reflecting the shifts of the basis vectors summed in subspaces P_0 and P_k $k = 1 : q$, respectively.

Substituting the expressions for $\Psi^{'}_{f_0}$, $\Psi^{''}_k$ and J_k, we derive:

$$a(x)\frac{1 + x^T}{\phi_{f^p_0}(x)} + \sum_{k=1}^{q} b_k(x)\frac{1 + x^T}{\phi_k(x)}\frac{c_k(x)}{\phi^*_{f^q_0}(x)} = 0, \tag{10}$$

and after the transformation we have:

$$a(x)\phi^*_{f^q_0}(x)\prod_{k=1}^{q}\phi_k(x) = \sum_{k=1}^{q} b_k(x)\phi_{f^q_0}(x)c_k(x)\prod_{s=1,s\neq k}^{q}\phi_s(x) \tag{11}$$

Consider the arbitrary k-th term in the right-hand side. There is no ϕ_k among its multipliers. However, this term should be dividable by ϕ_k since the left side of the equation and all the other terms in the right side contain this multiplier. By the proposition, ϕ_k is mutually simple with respect to ϕ_s $s \neq k$, with respect to c_k and ϕ_{f_0}, which means that with respect to $\phi_{f_0^q}$ as well.

To prove complete controllability, we should turn to a reversed system. It is formed by diverging chains, with one input and l outputs. The problem of its observability actually reduces to observability of the chain considered previously.

Theorem 5 *If system (4) is observable (controllable), then system (3) is γ_r-observable (γ_r-controllable).*

Theorem 6 *If a periodically non-stationary system S^f is γ_r-observable and γ_r-controllable, and all the dynamics matrices of this chain are nonsingular, then it is completely observable and controllable.*

Proof γ_r-observability of the chain allows determining, by its output, state x_i of the system at a time instant coinciding with the beginning of sequence γ_r. However, due to nonsingularity of all dynamics matrices, knowing x_i, it is possible to determine not only any subsequent state, but also any state preceding it, which indicates complete observability of the system. Using the principle of duality, a similar conclusion can be made with regard to controllability.

Example 2 We proceed to Example 1, specifying the type of the links. We use homogeneous chains, i.e., the ones that consist of similar links. In order to fulfill condition (4) from Theorem 4, for the chain links, we choose the dimension equal to 3 and for the merging link, the dimension equal to 2. Let us describe the links as follows:

– links of non-stationary chains:

$$f_{1,1} = f_{2,1} = f_{2,2} = f_{2,3} = \begin{bmatrix} 1 & 0 & 1 \\ 1 & 0 & 0 \\ 0 & 1 & 0 \end{bmatrix}, \quad g_{1,1} = g_{2,1} = g_{2,2} = g_{2,3} = \begin{bmatrix} 1 \\ 0 \\ 0 \end{bmatrix},$$

$$h_{1,1} = h_{2,1} = h_{2,2} = h_{2,3} = \begin{bmatrix} 0 & 0 & 1 \end{bmatrix},$$

$$\phi_{f_{1,1}} = \phi_{f_{2,1}} = \phi_{f_{2,2}} = \phi_{f_{2,3}} = x^3 + x^2 + 1,$$

– merging link of a non-stationary model:

$$f_0 = \begin{bmatrix} 1 & 1 \\ 1 & 0 \end{bmatrix}, \quad g_0 = \begin{bmatrix} 1 \\ 0 \end{bmatrix}, \quad h_0 = \begin{bmatrix} 0 & 1 \end{bmatrix}, \quad \phi_{f_0} = x^2 + x + 1$$

– merging links of stationary chains:

$$\overline{f}_{1,1} = \overline{f}_{2,1} = \overline{f}_{2,2} = \overline{f}_{2,3} = f_{1,1}, \quad \overline{g}_{1,1} = \overline{g}_{2,1} = \overline{g}_{2,2} = \overline{g}_{2,3} = g_{1,1},$$

$$\overline{h}_{2,2} = \overline{h}_{2,3} = \begin{bmatrix} 0 & 1 & 0 \end{bmatrix}, \quad \phi_{\overline{f}_{1,1}} = \phi_{\overline{f}_{2,1}} = \phi_{\overline{f}_{2,2}} = \phi_{\overline{f}_{2,3}} = x^3 + x^2 + 1,$$

– merging link of a stationary model:

$$\overline{f}_0 = f_0^2 = \begin{bmatrix} 0 & 1 \\ 1 & 1 \end{bmatrix}, \quad \overline{g}_0 = \begin{bmatrix} 1 & 1 \\ 1 & 0 \end{bmatrix}, \quad h_0 = \begin{bmatrix} 1 & 1 \end{bmatrix}, \quad \phi_{\overline{f}_0} = x^2 + x + 1.$$

All these links are observable and controllable, and their characteristic polynomials are irreducible, which guarantees the fulfillment of the conditions of Theorem 4 and, as a consequence, the conditions of Theorems 5 and 6.

5 Conclusion

This paper considers the synthesis of a diagnostic model of a distributed real-time computing system. The model is intended to be used for test diagnosis with a parallel model, when the model is integrated in the system and executed in parallel with the main software. The results obtained make it possible to design more rational versions of the model in terms of the amount of diagnostic information transmitted as compared with the models obtained using well-known synthesis algorithms. For the resulting models, sufficient conditions for observability and controllability have been formulated.

This work was supported by the Russian Foundation for Basic Research, project no. 19-08-00052.

References

1. Ben, L., Khlif-Bouassida, M., Toguyeni, A.: On-the-fly diagnosability analysis of bounded and unbounded labeled petri nets using verifier nets. Int. J. Appl. Math. Comput. Sci. **28**, 269–281 (2018)
2. Broekman, B., Notenboom, E.: Testing Embedded Software. Addison-Wesley (2002)
3. Cormen, T.H., Leiserson, C.E., Rivest, R.L., Stein, C.: Introduction to Algorithms, 3rd edn. The MIT Press (2009)

4. Gruzlikov, A.M., Kolesov, N.V., Lukoyanov, E.V.: Test-based diagnosis of distributed computer system using a time-varying model. In: 10th IFAC Symposium on Fault Detection, Supervision and Safety for Technical Processes (2018)
5. Gruzlikov, A.M., Kolesov, N.V., Lukoyanov, E.V.: Test-based diagnosis of faults in data exchange addressing in computer systems using parallel model. J. Comput. Syst. Sci. Int. **57**(3), 420–433 (2018)
6. Gruzlikov, A.M., Kolesov, N.V., Tolmacheva, M.V.: Event monitoring of parallel computations. Int. J. Appl. Math. Comput. Sci. **25** (2015)
7. Gruzlikov, A.M., Kolesov, N.V.: Discrete-event diagnostic model for a distributed computational system. Merging chains. Autom. Remote Control **78**(4), 682–688 (2017)
8. Isermann, R.: Fault-Diagnosis Applications: Model-Based Condition Monitoring Actuators, Drives, Machinery, Plants, Sensors, and Fault-tolerant Systems. Springer Publishing Company, Incorporated (2014)
9. Kaldmuae, A., Kotta, U., Shumsky, A., Zhirabok, A.: Measurement feedback disturbance decoupling in discrete-time nonlinear systems. Automatica **49**(9), 2887–2891 (2013)
10. Lee, D., Yannakakis, M.: Principles and methods of testing finite state machines: a survey. Proc. IEEE **84**, 1090–1123 (1996). https://doi.org/10.1109/5.533956
11. Patton, R.J., Frank, P.M., Clark, R.N.: Issues in Fault Diagnosis for Dynamic Systems. Springer-Verlag, London (2000)
12. Zaytoon, J., Lafortune, S.: Overview of fault diagnosis methods for discrete event systems. Ann. Rev. Control **37**(2), 308–320 (2013)
13. Zhirabok, A., Shumsky, A.: Fault diagnosis in nonlinear hybrid systems. Int. J. Appl. Math. Comput. Sci. **28**, 635–648 (2018)

Diagnostics of Rotary Vane Vacuum Pumps Using Signal Processing, Analysis and Clustering Methods

Pawel Łój and Wojciech Cholewa

Abstract Rotary vane vacuum pumps, which are usually driven by a three-phase electric motor, are widely used in various types of industries. Eccentric placed rotor in the cylinder, combine with centrifugal force, forces the vanes to move out and into the slots milled in the rotor, resulting in increase and decrease of the inter-vane volume, causing the air to be sucked into the pump. During the exploitation, there is a possibility of a mechanical damage to the vane, blocking it in its slot, and in some cases the vane can fall out of its groove, what leads to catastrophic failure. The article presents a new method for diagnostics of vanes, based on the observation and analysis of a pressure signal, validated by the clustering process of the recorded signals, showing the way to assess the state of the device without taking it out if the operation.

Keywords Vacuum pump · Signal analysis · Technical diagnostics

1 Introduction

Rotary vane vacuum pumps, in industrial plants, are the most commonly used to produce, low vacuum (down to 1 mbar) and medium vacuum (down to 0,001 mbar). They are used in various types of industries like pharmaceutical, food, chemical or even plastic processing. They are also used in pneumatic transport and in all kind of holding and moving operations of components during its production [1–3]. There a two different types of rotary vane vacuum pumps—dry-running, which use self-lubricated graphite vanes, and oil-lubricated [2–4]. This article is written basing on the research carried out on the oil-lubricated rotary vane vacuum pumps, but it can be

P. Łój (✉) · W. Cholewa
Silesian University of Technology, ul. Konarskiego 18A, 44-100 Gliwice, Poland
e-mail: pawel.loj@polsl.pl

W. Cholewa
e-mail: wojciech.cholewa@polsl.pl

© The Author(s), under exclusive license to Springer Nature Switzerland AG 2021
J. Korbicz et al. (eds.), *Advances in Diagnostics of Processes and Systems*,
Studies in Systems, Decision and Control 313,
https://doi.org/10.1007/978-3-030-58964-6_8

Table 1 Technical data of selected oil-lubricated rotary vacuum pumps manufactured by BUSCH company [2]

Model	Nominal pumping speed $[m^3/h]$	Ultimate pressure [mbar]	Nominal motor speed $[min^{-1}]$	Weight [kg]
R5 RA0100F	100	0,1	1500	73
R5 RA0302D	300	0,1	1500	74
R5 RA0750A	750	0,1	1200	670
R5 RA1000B	1000	0,3	1000	1000
R5 RA1600B	1600	0,3	1000	1330

assumed, due to the similar principle of operation, that the presented methodology, without major obstacles, can be also applied in dry pump diagnostics. Some overview technical data of selected pump models are presented in Table 1.

Rotary vane vacuum pumps, are usually built in the same way: electric motor permanently coupled with by the pump unit, which is flange-fitted to the oil-mist separator unit. A most important band in such devices, is the pumps unit consisting of a cylinder, with inlet and outlet valves, vanes, rotor and side covers with bearing supports. Electric motor generates torque, which is transmitted through the claw coupling to the rotor (with milled slots for the vanes), setting it into the rotation movement [2–4]. The eccentric arrangement of the rotor in the cylinder, in combination with the centrifugal force, forces the vanes to move out and into the slots, what causes regular increase and decrease of the volume between the vanes, leading to suction, compression end exhausting the air into the oil-mist separator unit. The principle of operation of rotary vane vacuum pumps is illustrated in Fig. 1.

Experience shows, that properly operated and serviced rotary vane vacuum pumps can work for many years without showing significant wear symptoms and decrease in the level of ultimate pressure [6, 7]. However even during normal exploitation, as a result of different random events, there may occur damages of the vanes.

Even minor vane damage or uncontrolled vane displacements, like falling out of its slot, causes them to brake, and such situations leads to serious damage, which cost of repair is similar to the price of a new device. The examples of such damages are showed in the Fig. 2.

2 Rotary Vane Vacuum Pumps Diagnostics and Maintenance

As it was said, so far the only frequently used method in diagnostic procedure of rotary vane vacuum pumps, is based on the measuring the maximum pressure generated by the device and compare it with the pressure declared by the manufacturer on the pump shift plate. And in case of a significant decrease in the level of the generated

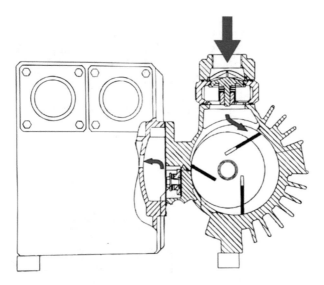

Fig. 1 Principle of operation of oil-lubricated rotary vane vacuum pumps

Fig. 2 Examples of damaged pumps due to vane malfunction [1]

vacuum, the pump is usually send to external company for service. This method has two major disadvantages. First of all, it does not allow assessing the condition of the device during operation, which involves the need to disconnect the pump from the cooperating machine or central vacuum system. Secondly, it only allows a binary assessment of the condition of the device being diagnosed, without any possibility of historical record and future predictions about the state of the pump during operation. To date, no diagnostic methods are known to identify progressive damages without shutting down the device [5].

The idea of a new diagnostic method, is based on the registration and analysis of the pressure signal, generated by the rotary vane vacuum pump, extracting statistical measures from it, and using them to conclude about the past, current and future state of the device.

The pressure value is measured at the pump inlet, directly in the suction port, by the analog pressure sensor, connected to suction port by the short hose (inner diameter $\phi2$

Fig. 3 Measuring system

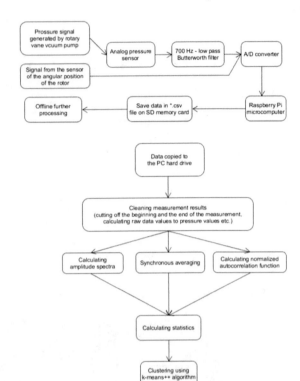

Fig. 4 The signal
processing path

mm and 50 mm long). Then the signal is filtered using a low-pass Butterworth filter, 700 Hz cutoff frequency. Filtered signal together with the signal pointing the angular position of the rotor, is sent to Raspberry Pi microcomputer with an analog-digital converter. All these measurements were carried out offline, which was necessary for reaserch and testing. Eventually, it is planned to perform an online diagnostic system, that realize both functions, measurements and calculations, by one microcomputer.

After completing the measurement, the data is saved on an external data carrier and transferred to a personal computer for further processing [5]. Diagram ilustarting the measurment system is presented in the Fig. 3, and the diagram ilustarting the signal processing path in presented in the Fig. 4.

3 Carried Out Research and Results

Research has been divided into two main phases. In the first phase, an experiment was carried out, involving the measurement of pressure on a device with simulated defects, in order to find statistical features that can carry information about the state of the device. Then, during the second phase of the research, in order to verify

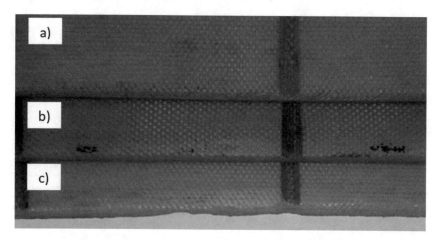

Fig. 5 Vanes used during the research—intact vane (**a**), vane with low degree of its defect (**b**), ane with high degree of its defect (**c**)

the method, a large-scale experiment was carried out, consisting on measuring and registration of the pressure signal generated by 36 devices in various states. Small-scale experimental studies were carried out on a test device, with a capacity of 100 m3/h, with low wear rate. The measurement of pressure was carried out with a sampling frequency 2048 Hz, using a low-pass Butterworth filter, 700 Hz cutoff frequency. The rotor rotational frequency was 25 Hz. During the tests, there was recorded the course of the pressure generated by the device, equipped with one damaged vane with low degree of its defect (a), with one damaged vane with high degree of its defect (b), with two damaged vanes with various degree its of defects (c), and equipped with undamaged vanes in conditions of lack of lubrication (d). Also the reference test, on the device with intact vanes, was performed (e). Vanes used it the reaserch are shown in the Fig. 5.

The pressure signal analysis was limited to time and frequency domain analysis. In the time domain, the synchronous averaging method was used, and also the value of the autocorrelation function was determined, while in the frequency domain the signal spectral analysis was used Fig. 6.

As a result of the synchronous averaging method, synchronized by the marked position of the rotor, an average time course of pressure was generated, for every individual state of the device. The coloured lines, which are shown in the Fig. 3a–e, present the pressure course during single rotations of rotor. Showing the whole measurement in this way, it allows to observe anomalies which occur during its time. It can be observed, that while the degree of the vane damage increases, the disorder of the chart also increase. In the case of a heavily damaged vane, two inflection points are clearly visible, and in the case of a slightly damaged vane, there are visible three inflection points, and the shape of the graph is similar to the shape of an undamaged device. This phenomenon is directly related to the suction through the edge of the

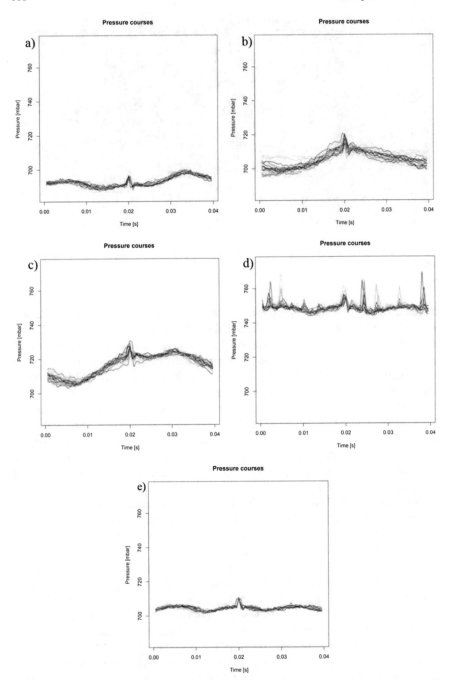

Fig. 6 Time courses of the generated ultimate pressure signals. Each coloured line presents a single rotation of the rotor— (see Sect. 3)

vane. In the case of insufficient lubrication, much more pressure peaks are visible, what is related to the discontinuity of the oil film between the cooperating elements.

By analysing the signal autocorrelation function graph, showed in the Fig. 7a–e, it is possible to conclude about the condition of the vanes working in a vacuum pump. In the case of even small damage, a strong autocorrelation can be seen only once per revolution, which in this measurement conditions is 82 samples long. With the growing degree of damage to a single vane, the strength of this autocorrelation decreases. In the case of efficient pumping, strong autocorrelation is visible 3 times per revolution, which is associated with the same operating conditions for all three vanes. In a situation where there is a lack of lubrication in the pump unit, a strong autocorrelation occurs also 3 times per revolution, however, the shape of the autocorrelogram envelope is chaotic.

By analysis of the amplitude spectra of the pressure signals, generated by the rotary vane vacuum pump, shown in the Fig. 8a–e, it is also possible to detect damages of the vanes working in it. In the spectrum of the pressure, generated by an efficient pump, the frequency band of 75 Hz dominates, which is related to the rotational frequency (25 Hz) of the rotor and the number of 3 rotating vanes. As the vane damage increases, the frequency band 25 Hz increase, which corresponds to the frequency at which the damaged vane passes by the suction port. In case of the insufficient lubrication, the pressure signal spectrum shows a significant increase in share of high-frequency components (150 Hz), associated with the operation of moving elements (such as inlet and outlet valves), which are not directly related to the rotational frequency of the rotor.

Based on the results of the measurements and observations of the time pressure coures, pressure spectrum charts and autocorrelograms of the pressure signal, 25 statistical features were calculated, what reduced the presented charts to numerical data. Then statictical features, which have a significant impact on the absolute condition of the device were chosen. The selected statictical measures are presented in Tables 2 and 3.

From considered statistical point features, it follows that:

– Any damage causes an increase in the standard deviation.
– Problems associated with the lubrication system, can be detected by analysing the kurtosis of the obtained time pressure course.
– Even a small damage of the vane, is almost immediately visible in the change of the value of the normalized autocorrelation function of the pressure time course.
– With the increase of the blade damage, the spectral power and the share of the spectral power of the frequency band, corresponding to pressure changes once and twice per rotor rotation, also increases.
– The share of the spectral power of a high frequency band, increase while the lubrication problems occurs.

Further work was based on carrying out a large-scale experiment, consisting in recording the pressure signal, generated by 36 different rotary vane vacuum pumps, used in various production and processing plants, with different history and at different conditions. On all these measurements, the same calculation process was carried

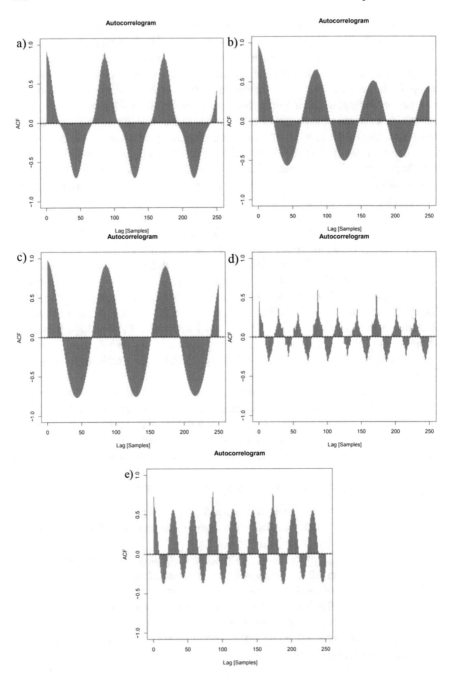

Fig. 7 Graphs of the normalized autocorrelation function of the pressure signals

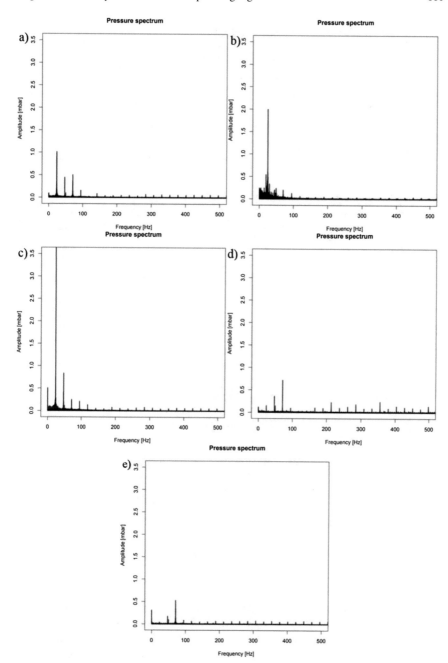

Fig. 8 Amplitude spectra of the generated pressure signals

Table 2 Calculated spectral power in frequency bands

ID (see Sect. 3)	Kurtosis	Standard deviation	Autocorrelation value for 1 revolution	Autocorrelation value for 1/2 revolution	Autocorrelation value for 1/3 revolution
a	−1.06	5.91	0.93	−0.80	−0.58
b	−0.48	2.66	0.89	−0.68	−0.34
c	−0.43	4.91	0.87	−0.59	−0.43
d	10.60	2.12	0.57	−0.35	0.35
e	2.87	1.43	0.78	−0.56	0.56

Table 3 Calculated spectral power in frequency bands

ID (see Sect. 3)	25 (±8) Hz band signal power [%]	50 (±8) Hz band signal power [%]	75 (±8) Hz band signal power [%]	<150 Hz band signal power [%]	Signal power [mbar²]
a	87.8	6.3	1.0	1.4	1.18
b	68.1	6.7	14.1	5.8	0.23
c	72.5	5.2	1.5	2.6	0.75
d	2.7	10.4	33.5	42.7	0.12
e	1.2	6.5	53.3	22.0	0.06

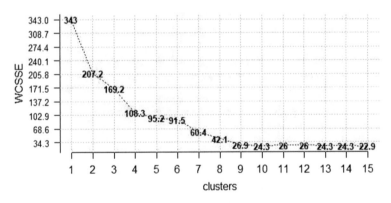

Fig. 9 Value of WCSSE (within cluster sum of squared error) for numer of clusters

out, and basing on signals features from Tables 2 and 3 (the kurtosis and standard deviation were rejected because they disturbed the clustering process), measured devices were subjected to the clustering process. The clustering process was based on the k-means++ algorithm, and the number of number of clusters was selected based on the chart show in the Fig. 9 [8].

The k-means++ algorithm has been chosen for its easy implementation and for its sensitivity to singular points. It can be assumed that due to the incompleteness of the learning set, each device, with a high degree its defects, will be a kind of

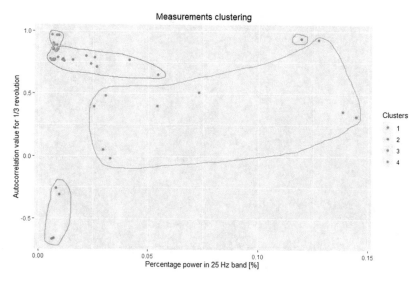

Fig. 10 Generated clusters

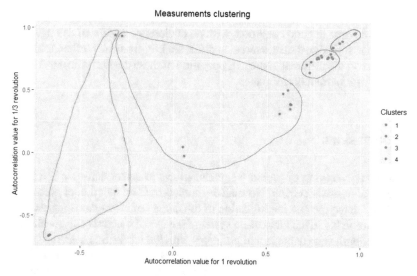

Fig. 11 Generated clusters

singularity in every attribute space. Nevertheless, it will be located relatively close to the cluster containing damaged devices which means that the k-means++ algorithm should qualify such a device to the appropriate cluster [8, 9].

The results of the grouping process are shown in Figs. 10, 11, 12. There are 4 clusters of the devices. Cluster 1 and cluster 4 consist of devices with alarming signal parameters, in which advanced and progressive damage can be suspected. Cluster 2

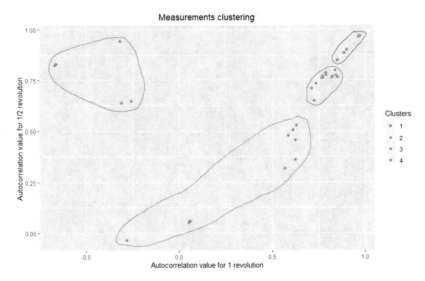

Fig. 12 Generated clusters

consists of devices in good condition, with no obvious symptoms of damage, whose values of pressure signal characteristics differ slightly from each other, which may have a cause in the different history of operation of these devices. Cluster 3 consists of devices with perfect performance.

4 Conclusions

Synchronous averaging of pressure signals, allows to detect damages and identify the degree of its advancement. Normalized autocorrelation function of pressure signals, allows detection and identification of the degree of vanes damages, as well as problems occurring in the lubrication system. The use of spectral analysis allows the detection of both vane damage and problems with the lubrication system.

The analysis of the presented grouping process shows that the pressure signal features mentioned above can be taken into account to determine the state of the rotary vane vacuum pump, which makes them a valuable and significant diagnostic features.

The clustering analysis proves the relationship between the condition of the device and the selected signal features. It allows for the next step of investigations associated with the construction of an automatic classifier and a system that diagnoses rotary vane vacuum pumps during exploitation. The basic idea of the automatic online diagnosing system for rotary vane vacuum pumps is presentet in the Fig. 13.

Fig. 13 The idea of the online diagnostic system

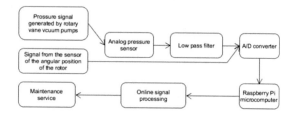

References

1. Łój, P.: Exploitation of rotary vane vacuum pumps. In: Modelowanie Inżynierskie, vol. 38, no. 69. pp. 56–59 (2018)
2. Busch.: wwww.buschvacuum.com/pl/pl. Accessed 05 April 2020
3. Becker.: www.becker-polska.com. Accessed 19 Sep 2019
4. Pfeifer Vacuum GmBH: The Vacuum Technology Book. Pfeifer Vacuum GmBH, Aßlar (2013)
5. Łój, P., Cholewa, W.: Diagnostics of rotary vane vacuum pumps using signal analysis and processing methods. In: Vibrations in Physical Systems, 2019, vol. 30, no. 2 (2019)
6. Hałas, A. : Technika próżni, Oficyna wydawnicza Politechniki Wrocławskiej, Wrocław (2017)
7. Yoshimura, N.: Vacuum Technology. Springer, Berlin Heidelberg, Berlin (2008)
8. Chalouli, M., Berrached, N., Denai, M.: Intelligent health monitoring of machine bearings based on feature extraction. J. Fail. Anal. Preven. **17**, 1053–1066 (2017)
9. Fan, L., Chai, Y., Li, Y., (2020) A density-based k-means++ algorithm for imbalanced datasets clustering. In: Jia Y., Du J., Zhang W. (eds) Proceedings of 2019 Chinese Intelligent Systems Conference. CISC: Lecture Notes in Electrical Engineering, vol. 594. Springer, Singapore (2019)

Neural Modelling of Steam Turbine Control Stage

Jerzy Głuch and Marta Drosińska-Komor

Abstract The paper describes possibility of steam turbine control stage neural model creation. It is of great importance because wider application of green energy causes severe conditions for control of energy generation systems operation Results of chosen steam turbine of 200 MW power measurements are applied as an example showing way of neural model creation. They serve as training and testing data of such neural model. Relatively simple set of nozzle boxes neural models consisting control stage is applied. They act as a neural regresor. Research study on ways of creation mentioned neural model is the main purpose of the paper. Finally accurate neural tool is created. It can serve as a proper pattern of control stage operation for engineers tuning turbine control equipment. Another way of application consists in usage as a component of turbomachinery heat and flow diagnostic programs. These programs take mainly into account of technical objects efficiency degradation.

Keywords Control stage · Steam turbine · Artificial neural networks · Diagnostics

1 Introduction

Large power steam turbines installed in power stations are operating under variable load and variable ambient conditions [1]. This necessitates the use of precise control, which play an important role in the process of safe and correct use of steam turbines in order to operate at the best possible efficiency [2–5]. This is associated with financial profits for energy generation [6, 7]. The adjustment process uses control valves [8, 9] driven by various servo motors. Power control of the power facility requires variations of steam flow rate through the steam turbine to be adjusted [10, 11]. So steam flow is increased or reduced here. The adjustment process, i.e. changes in the flow rate of the working medium (in this case water steam) can be carried out by various methods listed below [9]:

J. Głuch (✉) · M. Drosińska-Komor
Faculty of Ocean Engineering and Ship Technology, Gdańsk University of Technology,
Narutowicza 11/12, 80-233 Gdańsk, Poland
e-mail: jgluch@pg.edu.pl

© The Author(s), under exclusive license to Springer Nature Switzerland AG 2021   117
J. Korbicz et al. (eds.), *Advances in Diagnostics of Processes and Systems*,
Studies in Systems, Decision and Control 313,
https://doi.org/10.1007/978-3-030-58964-6_9

- Throttle governing (quality control);
- Quantity governing;
- By-pass governing;
- Follow-up control.

Throttling adjustment is carried out by several cooperating valves. This type of control applies to low-power turbines, ship turbines and those where partial loads operations are very rare.

The quantity control is designed for medium-power turbines where load changes occur very often. The steam are delivered to the turbine by nozzle boxes where each single box is equipped with its own control valve. In the case of partial load of the turbine, some of the valves are partially open and the others are fully open or closed. This control usually requires application of a turbine control stage in the turbine which is feeded on the part of the circuit blade system [12].

Bypass control is applied when a maximum steam pressure at the inlet of the first stage of the turbine occurs. In this case, the flow rate present in the turbine depends on the surface of the first stage nozzles [13]. This control applies to small-power turbines or to some nuclear power stations.

Follow-up control can be carried out by valve cooperating with pressure changes before and after the boiler using a variable-capacity feed water pump. Due to the highest efficiency among the methods of control, it is used in large power turbine cycles.

In following parts of the paper attention will be focused on the possibility of modelling of the quantity control. The purpose of the paper is to show a possibility of such modelling creation and application to numerical simulations and diagnostics of the steam turbines energy transfer efficiency. The modelling method will be presented on the example of a 200 MW turbine power unit which is the most popular in the Polish energy supply system.

Physical model of steam turbine cycle components and their connections is presented in Fig. 1. Way of control has a great influence on distribution of all flow parameters which in turn decide about energy transfer processes efficiency [1, 6].

Control valves are placed before turbine High Pressure body. They operate together with the first turbine stage. The entire control unit is marked with an ellipse in Fig. 1. Its physical model is very complex. It is demanded that its operation should be done by a calculation model properly performing its characteristics. It is important both for the design and for exploitation parameters restoration. The second demand applies mainly for diagnostic s purposes of energy transition processes. The authors are members of the research team modelling the thermal cycles of steam turbines. Performance of iterative sequential models occurs the best. They involve the inclusion of individual component's balances of energy, mass and momentum to the calculation and simulation of the cycle operation. Individual calculation results shall be transfer further according to the structure of the cycle. For computational purposes, the cycle structure is modelled by graph, Fig. 2. This model has already been described many times, e.g. [14, 15]. The main equations of convergence are solved in the control stage model, establishing computational distributions of mass, energy and pressure.

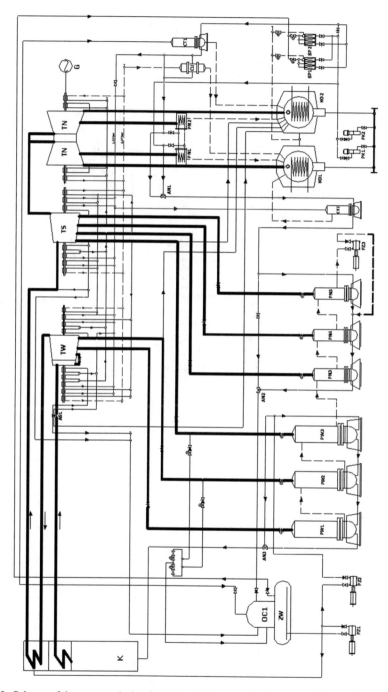

Fig. 1 Scheme of the power unit showing place of HP control valves marked by an ellipse

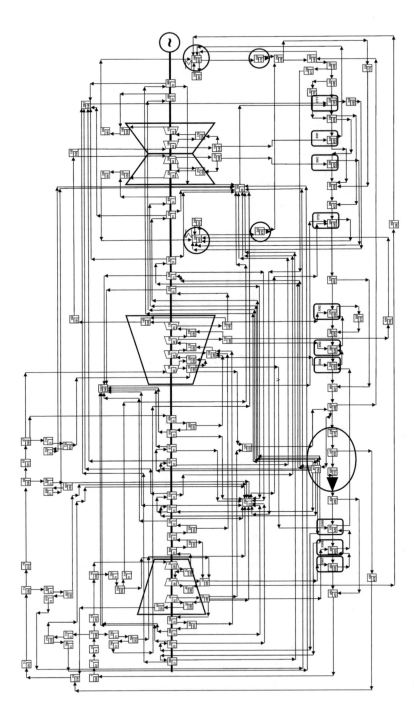

Fig. 2 Numerical model—graph—of the power unit showing place of HP control valves and control stage marked by an ellipse. Scheme built on the basis of Fig. 1. Each of rectangle stands for unique cycle component calculation procedure. Lines stand for the cycle pipelines

The accuracy of the calculation depends mainly on the accuracy of the modelling of control stage efficiency. The algebraic-physical calculation alone is not enough due to the significant dependence of the control stage performance on the way of operation. Practically every turbine has a different characteristics of its control stage. Therefore, its performance resulting in changes of efficiency, should be tuned individually to operating conditions on the basis of different safety measurements. Presentation of neural modelling of control stage leading to an accurate determination of its efficiency is the main goal of the paper. The paper presents the results of research studies about usage of artificial neural networks (ANN) for this purpose.

2 Object of Investigations

In this paper, attention is focused on a 200 MW steam turbine installed in one of the Polish power plants. The turbine is a three-body unit: high pressure body HP, intermediate pressure body IP and low pressure bodies LP (Fig. 1). In the considered cycle, 7 regenerative heat exchangers are used to raise the temperature of the working fluid before the boiler. Regenerative exchangers are heated by steam taken from extractions of the turbine. The system also includes a deaeration component which is a special type of heat exchangers. Secondary superheating is also used in the system. These solutions improve the safety of the operation of the power unit and better efficiency of turbines operating in the cycle.

Control valves CV are installed at the intake steam pipelines of the turbine before HP body and IP body. Control valves for IP body are used only at the start-up and shut-down of the turbine and, despite their name, do not play a control role. HP control valves play this role and they are used for the process of flow control of the working fluid. The place of their installation is marked with an ellipse in the Fig. 1 and its numerical graph, Fig. 2.

HP control valves have only recently been powered by mechanical servo motors. Currently, they are powered by servo motors of the electrohydraulic system EHR, which allow to change the settings of the controllers even during inter-reparation periods. Hence, the fast and precise modelling of these valves operation plays an important role in exploitation procedures. It can be also used to diagnose both the condition of this cycle component and the controlled turbine in terms of the efficiency of the energy transfer.

The system of HP control valves is shown in a fragment the piping diagram for the HP turbine body (Fig. 3) describing them with the symbol CV. There exist 4 CVs, so control stage consists of 4 nozzle boxes.

3 Analytical Calculation Model

Control valves direct steam into a single nozzles group. Each group is characterized by a different flow field A and its valve is opened individually at power much lower

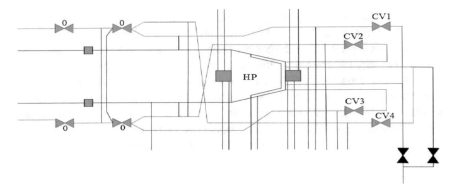

Fig. 3 Distribution of pipelines and control valves CV in a HP turbine body

than the design power. Valve CV4 is closed first, later CV3 and from this moment they are completely closed while the others are opened to varying degrees. Therefore, there is a different p_{in} pressure before each nozzle group. Behind each nozzle group there is the same pressure p_{out} called the pressure at the control stage chamber. Hence, the mass flow rates by an individual group of degrees can be described by an equation:

$$m_j = f(A_j, \Delta p_j), \Delta p = p_{in} - p_{out} \tag{1}$$

where:

m—mass flow rate,
A—outlet cross-section of a nozzle box,
Δp—pressure drop in a nozzle box,
j—number of a nozzle box.

Total mass flow rate m_{tot} of the steam flowing by control stage treated together with control valves is equal:

$$m_{tot} = \sum m_j \tag{2}$$

Mass flow distribution is necessary to calculate energetical efficiency of the control stage. Knowledge of the values of enthalpy obtained on the basis of local temperatures and pressures is also necessary. For the purpose of this paper they are obtained from thermal and flow measurements in the turbine cycle. Thus applied general efficiency η equation is:

$$\eta_{tot} = \frac{\sum m_j(h_{inj} - h_{outj})}{m_{tot}(h_{intot} - h_{outstot})} \tag{3}$$

where:

h_{in}—inlet enthalpy—$h_{in} = f(p_{in}, t_{in})$,
h_{out}—outlet enthalpy—$h_{out} = f(p_{out}, t_{out})$,
f—generally stands for a function,
j—number of a nozzle box,
tot—concern total control stage,
s—concern isentropic (ideal) expansion.

Manufactured turbine has already constant geometric parameters, so to the control stage cross-section for nozzle groups are already fixed. The inlet and outlet pressures of each nozzle group, which depend on the turbine load parameters, called independent analytical model parameters, shall be changed. It was found in [15] that for steam power units of 200 MW there are 8 such parameters: live steam mass rate, live steam pressure and temperature, secondary steam temperature, condensation pressure dependent on the temperature of the condenser cooling water, deaeration pressure and water injection flows to live steam superheater and to secondary superheater.

The quantities in Eqs. (1), (2) and (3) are also functions of mentioned independent parameters. Nevertheless, the depiction of the algebraic character is complicated and difficult. An additional difficulty consists in obtaining accurate results of calculations in relation to the values actually measured on the power unit [15]. Hence, research has been performed to prepare proper control stage model using artificial neural networks ANN regresor type.

4 Control Stage Experimental Modelling

The studies carried out led to the presentation of the neural model as a many variables function. In the practice of running power units, the main requirements of the manufacturers of the main circuit components (the boiler and turbine) must be fulfilled. This results in very limited variability of many independent parameters. However, when the required limits are exceeded, mathematical errors due to extrapolation. Data from previous studies of authors presented in [15] and the tests performed led to a reduction of these variables to two quantities: a live steam flow rate m_p and the temperature of the condensers cooling water t_w.

Further research used the results of measurements of one of a power unit during the winter-spring–summer period to build a neural model. A set of four neural models was created to determine the pressure before the control valves and one determining pressure after the nozzle groups in the control stage chamber. Some characteristics of the measurement data used for SSN training are shown in Figs. 4 and 5.

Used experimental data have been obtained from steam turbine DCS system and they have been collected for several days. It is possible to obtain 234 thermal and flow data from this particular DCS measured practically at the same moment of time, creating one elementary measuring set. Each 10 s new measured data set has been obtained (more than 10,000 sets for one day). From these data the most quiet operating subsets (almost constant turbine power) have been chosen, each containing

Experimental dependence N = f(mp,tw)

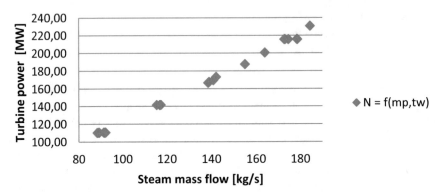

Fig. 4 Experimental dependence of the power on live steam flow and condenser cooling water temperature

measured pCVj = f(mp,tw)

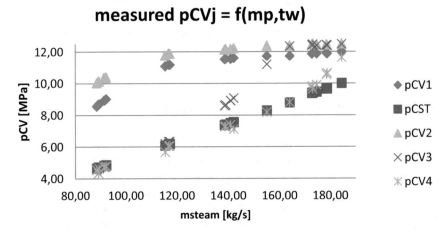

Fig. 5 Experimental dependence of the pressure behind control valves and in control stage chamber on live steam flow and condenser cooling water temperature

more than 100 elementary sets. Finally 27 such subsets have been obtained. They are characterised by different turbine power and different cooling water temperature. Mean values of thermal and flow measured parameters have been calculated in each subset. The action minimise potential measurement errors for real DCS systems for the purpose of the paper. All necessary measurements of control stage pressures and temperatures are among these measurement data.

The used experimental data was collected in function of variability of the live steam mass rate of 88 to 184 kg/s, which correspond to the turbine load of 50 to 100% of the design power and for the condensers cooling water temperature of 3 to 24 °C. In practice, they cover the entire operating range of the turbine after its thermal stabilization.

Different simple neural network structures have been chosen for research and tested. Finding proper nozzle box inlet pressure is the main task of tested ANN. Tests showed finally that 2 neurons in the hidden layer are enough to build each of the neural model consisted of: 2 inputs m_p and t_w and one output—the searched pressure [15, 17]. The Levenberg–Marquardt method for ANN training was selected because it gave good results in previous similar authors research [15–17].

5 Performance Characteristics and Application of Neural Models

The pressure curve forms determined by the neural models of the control valves and by the control stage chamber are presented in Figs. 6 and 7. Figure 6 shows influence of the load of the power unit onto moments of opening selected control valves.

Figure 7 is based on the results of the neural calculations presented in Fig. 6 and can be used to determine the mass streams flowing through the different nozzle groups according to Eqs. (1) and (2).

The use of the results of the neural calculations of the control stage can be applied in two ways.

The first possibility to use the results of the neural calculations of the control stage is the usage of diagrams of type shown in Fig. 6. They assure checking the settings of hydraulic servo motors for electrohydraulic control EHR. The purpose is to establish the highest possible efficiency of the control stage operating under various loads. As mentioned, this modern control can be carried out relatively simply even during

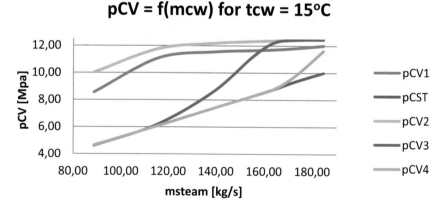

Fig. 6 An example of neural dependence of the pressure behind control valves and in control stage chamber on live steam flow and given condenser cooling water temperature

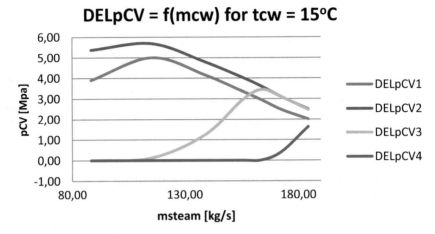

Fig. 7 An example of dependence of the pressure drop in nozzle boxes described from their neural models and possible to apply to determine mass flow rates distributions for nozzle boxes as function of live steam flow and condenser cooling water temperature

inter-repair periods. In addition to efficiency, the setting have significant impact on the level of vibration of the turbomachinery. It can therefore be an auxiliary model for setting the controllers that exist on SCADA or DCS systems. Sometimes it can serve as a pattern of proper operation.

The second possibility to use consists in the use of the results of neural calculations such as those presented in Fig. 7. Steam mass rates of each of the nozzle boxes may be determined in function of the specified total live steam flow rate on the nozzle groups determining neurally the pressure differences resulting from pressure drop. In the next step, the energy balance of the control stage can be done, The average temperature in the control stage chamber can be also determined neurally applying also Eqs. (1) and (2).

Finally on this basis its efficiency and the control stage developed power can be determined. The distributions of these parameters are presented in Figs. 8 and 9. It is particularly interesting to use Fig. 8. It may be used to establish a pattern of efficient control operation and hence to detect operational degradation in the control stage of the steam turbine.

One can also plan usage of neural calculation in order to apply a neural model of control stage as a diagnostic component in existing steam power units thermal and flow diagnostic programs, e.g. [16].

6　Summary

The paper provides a method for neural modelling of steam turbines control stages. The results of measurements of one of the steam turbines in Polish power plants were

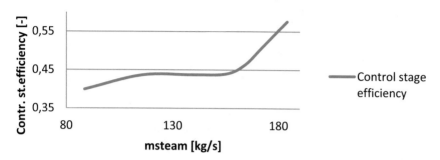

Fig. 8 An example of the dependence of control stage efficiency on the live steam flow rate of power unit for given condenser cooling water temperature of 15 °C

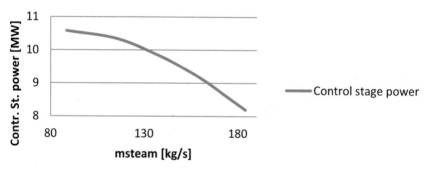

Fig. 9 An example of the dependence of control stage power on the live steam flow rate of power unit for given condenser cooling water temperature of 15 °C

used to train artificial neural networks. Simple ANN was used to achieve satisfactory accuracy of control stage neural calculations as well as a simple and fast training. The neural model can be used to check the settings of hydraulic servo motors when adjusting electrohydraulic systems EHR. It can also be used as a component of steam power units thermal and flow diagnostics programs.

References

1. Szewczuk-Krypa, N., Drosińska-Komor, M., Głuch, J., Breńkacz, Ł.: Comparison analysis of selected nuclear power plants supplied with helium from high-temperature gas-cooled reactor. Polish Maritime Res. **25**(s1), 204–210 (2018)

2. Breńkacz, Ł., Żywica, G., Drosińska-Komor, M.: The experimental identification of the dynamic coefficients of two hydrodynamic journal bearings operating at constant rotational speed and under nonlinear conditions. Polish Maritime Res. **24**(4/96), 108–115 (2017)
3. Szewczuk-Krypa, N., Grzymkowska, A., Głuch, J.: Comparative analysis of thermodynamic cycles of selected nuclear ship power plants with high-temperature helium-cooled nuclear reactor. Polish Maritime Res. **25**(s1), 218–224 (2018)
4. Breńkacz, Ł., Żywica, G., Drosińska-Komor, M., Szewczuk-Krypa, N.: The experimental determination of bearings dynamic coefficients in a wide range of rotational speeds, taking into account the resonance and hydrodynamic instability. In: Awrejcewicz J. (eds) Springer Proceedings in Mathematics and Statistics Dynamical Systems in Applications. DSTA 2017, vol. 249, pp. 13–24. Springer, Cham (2018)
5. Głuch, S., Piwowarski M.: Enhanced Master cycle—significant improvement of steam rankine cycle. In: 25th International Conference in Engineering Mechanics, pp. 125–28, Svratka, Czechia (2019)
6. Windemuth, C., Lange M., Mailach R.: Introduction of a novel test rig for the investigation of fluid-structure interaction effects in steam turbine control valves using an elastic model. In: 13th European Turbomachinery Conference on Turbomachinery Fluid Dynamics and Thermodynamics, pp. 1–11. Lausanne, Switzerland (2019)
7. Błaszczyk, A., Głuch, J., Gardzilewicz, A.: Operating and economic conditions of cooling water control for marine steam turbine condensers. Polish Maritime Res. **18**(3), 48–54 (2011)
8. Kollross P., Tajč L.: An experimental research of the DSPW steam turbine control valve. In: 18th Conference of Power System Engineering, Thermodynamics and Fluid Mechanics AIP Conference Proceedings, 2189, pp. 020012-1–020012-8 (2019)
9. Wang, P., Liu, Y.: Unsteady flow behavior of a steam turbine control valve in the choked condition: field measurement, detached eddy simulation and acoustic modal analysis. Appl. Therm. Eng. **117**, 725–739 (2017)
10. Wang, P., Ma, H., Liu, Y.: Proper orthogonal decomposition and extended-proper orthogonal decomposition analysis of pressure fluctuations and vortex structures inside a steam turbine control valve. J. Eng. Gas Turbines Power **141**(4), 041035 (2019)
11. Pondini, M., Signorini, A., Colla, V.: Steam turbine control valve and actuation system modeling for dynamics analysis. Energy Procedia **105**, 1651–1656 (2017)
12. Wang, P., Ma, H., Quay, B., Santavicca, D.A., Liu, Y.: Computational fluid dynamics of steam flow in a turbine control valve with a bell-shaped spindle. Appl. Therm. Eng. **129**, 1333–1347 (2018)
13. Domachowski, Z.: Automatic Control of Thermal Turbosets (in Polish). Gdańsk University of Technology Publishers, Gdańsk (2011)
14. Głuch, J.: On Application of Neural Simulators Through Steam Turbines Blading Systems to Reference State Determination in Thermal and Flow Diagnostics (in Polish). Gdańsk University of Technology Publishers, Gdańsk (2014)
15. Głuch, J.: Selected problems of determining an efficient operation standard in contemporary heat-and-flow diagnostics. Polish Maritime Res. **16**(S1(36)), 22–28 (2009)
16. Gardzilewicz, A., Głuch, J., Bogulicz, M., Walkowiak, R., Najwer M., Kiebdój J.: Experience in application of thermal diagnostics in the turow power station. (CD-ROM) ASME IJPGC'2003, IJPGC2003–40017 (2003)
17. Głuch, J., Ślęzak-Żołna, J.: Solving problems with patterns for heat and flow diagnostics dedicated for turbine power plants. In: Int Gas Turbine Inst Proceedings Of The ASME Turbo Expo 2012. pp. 969–979 (2012)

Diagnostic of Calf's Body Temperature by Using Thermal Imaging Camera and Correction of Camera Errors

Wojciech Rafajłowicz⊙, Anna Rząsa⊙, and Paulina Jawor⊙

Abstract In this paper, we present a method of using a thermal imaging camera to measure the external temperature of a calf. We select the eye as the best spot that can be used to assess an animal's internal temperature. We also show a method of dealing with the nonstationarity of a low-cost thermal imaging camera.

Keywords Veterinary · Image processing · Thermal imaging · Contact-less temperature measurement

1 Introduction

Body temperature of animals is a very important indicator of their health, physiological status and welfare. Theoretically, measurement of internal temperature is a basic tool in clinical treatment. In practice, especially in farm animals, it doesn't work as expected because of physical effort and time on the one hand, the requirement to immobilize the animal on the other. Additionally, the stress of handling can result in an increase of temperature which interferes with the correct interpretation of the correct results [5, 8]. Looking for other non-invasive, precise, reliable and efficient methods of temperature measurement in animals is a challenge. Infrared technology looks very promising in this area and an infrared camera alongside scintigraphy, x-ray and ultrasound is recommended as a diagnostic tool in livestock animals [1, 10, 13]. Accurate measurement of bovine body temperature in a non-contact way is difficult due to its thick body hair. But it is widely reported that the eye, nose, vulvar and udder surface temperature are good substitutions for rectal temperature.

W. Rafajłowicz (✉)
Wrocław University of Science and Technology, Wybrzeże Wyspiańskiego 27, 50 370 Wrocław, Poland
e-mail: wojciech.rafajlowicz@pwr.edu.pl

A. Rząsa · P. Jawor
Department of Immunology, Pathophysiology and Veterinary Prevention, Wrocław University of Environmental and Life Sciences, Wrocław, Poland

© The Author(s), under exclusive license to Springer Nature Switzerland AG 2021
J. Korbicz et al. (eds.), *Advances in Diagnostics of Processes and Systems*, Studies in Systems, Decision and Control 313, https://doi.org/10.1007/978-3-030-58964-6_10

It looks very promising in routine husbandry animal health monitoring [3, 4, 11, 13, 14]. The aim of the study was to evaluate the correlations between rectal (traditional thermometer) and eye/ear (infrared camera) temperature in calves. The study was carried out on 81 calves between 9 days and 3 months old from three herds (A,B,C).

The outline of the paper is as follows: firstly we will present the basics of thermal imaging. Then our problem will be stated. An image processing solution will be shown. Finally, a problem of correcting error due to nonstationarity of measurements will be presented and solved.

2 IR Cameras

In recent years thermal imaging cameras have shifted from being expensive and rare into small affordable devices with multiple uses. Not only have prices dropped, but, thermal cameras (commonly called IR cameras for short) are much easier to use and do not require separate cooling. Medical imaging is one of many uses, but we should not forget about other uses, including electrical maintenance, assessing a building's thermal properties and many others. Especially the requirement for energy efficiency certification is the main reason for the proliferation of IR cameras.

Most image analysis is carried out by the human eye with just simple software tools. Most of them allow us only to change colors attributed to different temperatures, changing the temperature range and see the temperature profile along a specified line. Usually, an even more limited number of functions are available in the camera firmware.

General interpretation of a thermal image requires some experience. Additionally, any form of automatic processing would be tailored only to a specific problem. In this paper, we want to present one such solution in the area of veterinary diagnostics. We will also present how to solve some drawbacks of popular, low-cost thermal imaging cameras.

3 Basics of Thermal Imaging

Every object in a temperature above absolute zero (0K) spontaneously emits some energy in the electromagnetic spectrum. This emission is continuous. Its distribution is given by Planck's law for an ideal black body.

$$B_\lambda(\lambda, T) = \frac{2hc^2}{\lambda^5} \frac{1}{e^{\frac{hc}{\lambda k_B T}} - 1}. \tag{1}$$

Generally, only in high temperatures do objects emit light in the visual spectrum. Below (in the sense of frequency) this range lies within the infrared band where what is commonly called heat is being emitted in a typical temperature range.

This allows for many methods of contactless temperature measurements. The simplest one is useful in case when radiated energy is in the visible spectrum – as in the case of glowing metal. The method is called a disappearing-filament pyrometer.

By using special lenses (normal and optical glass is reflective to IR) emitted radiation is focused on some kind of detective device. Generally, these detectors are of the bolometric type where every pixel is heated by thermal radiation, its temperature measured and then the temperature of the object is being calculated. This process causes an increase in temperature of the sensor itself and can, in turn, falsify the reading.

The amount of emitted energy depends not only on temperature but also on the emissivity. Generally it is taken into account as a coefficient valued 1. for ideal black body (full theoretical emission) and 0. for ideal white body (which reflects everything but does not emit).

4　Image Processing Problem

One common veterinary procedure is measuring an animal's internal temperature. Using a traditional, contact thermometer has many disadvantages. Mainly it takes time and a live subject can not be expected to cooperate. Thermal imaging can prove to be the answer to that problem. Before this, certain problems should be solved. Most of them are traditional in image processing.

- Image acquisition—a thermographics image has to be created,
- Choosing the measurement spot—most of the animal is covered by its coat or fur, and whose main function is preventing heat loss by thermal isolation. Therefore, the animal's eye was chosen,
- Image segmentation—an eye has to be detected,
- Temperature measurement,
- Temperature correction—eye temperature is not an internal animal temperature due to different emissivity, also errors of the camera should be corrected.

5　Proposed Image Processing

In this section image processing methods are described (see [12]). Such methods are widely used in veterinary and medicine like in [2, 6].

5.1 Image Acquisition

Currently, due to the high costs associated with their purchase, the permanent installation of an IR camera in places where animals are kept is impractical. The measurement had to be taken using a hand-held camera in conditions far from optimal conditions. In addition, the camera is running changes. Typically, it was left running from 200 to 500 s. Since heat accumulation in a bolometric matrix does not stop, the time at which the image is taken is important. Fortunately, most electronic devices are able to record this.

5.2 Choosing Measurement Spot

An eye was chosen as a trial spot for measurements due to being visible at all times. It can also be presumed to have a temperature less dependent on the environment and more on internal body temperature.

Another reason for choosing an eye is the fact that its temperature is the highest. It makes it easy to find in the image by image processing methods. The relatively high temperature also suggests that this point is the closest to the internal temperature of the animal (Figs. 1, 2 and 3).

Steps carried out during processing our thermal image follow traditional patterns in image processing:

– Image segmentation—thresholding,
– Detecting eye—the Hough transform.

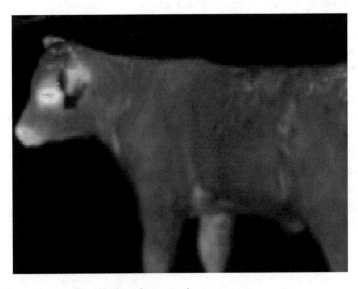

Fig. 1 Thermal image of a calf—base for processing

Fig. 2 Thermal image after thresholding

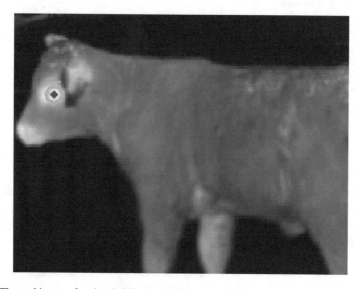

Fig. 3 Thermal image after thresholding and with detected circle

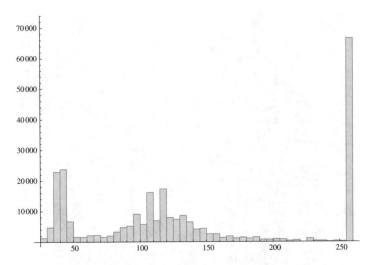

Fig. 4 A histogram of gray levels for choosing the correct threshold

5.3 Thresholding

Thresholding is probably the most common segmentation method. In the simplest variant, known as binary thresholding, it works as a simple transformation applied to each point of the image (point transformation). If its value is larger than the specified threshold then the result is 1 and 0 otherwise.

$$\forall_{x,y \in I} = \begin{cases} 1 \text{ if } I(x, y) \geq threshold, \\ 0 \text{ if } I(x, y) < threshold. \end{cases} \tag{2}$$

Other, more advanced methods of thresholding were devised but their use was not required in this problem.

The main problem in thresholding is a selection of the correct threshold level. Generally, the simplest method is to use the histogram of an image. The amount of each value (proportional to temperature) can be helpful to determine a good threshold. The histogram of the calf's thermogram is shown in Fig. 4.

We can clearly see that selecting the highest temperature area is relatively easy. In Fig. 1 we can see that the eye is probably the hottest point in the image. We choose the threshold accordingly resulting in the image in Fig. 2.

5.4 Detecting Eye Position

In the thresholded image (see Fig. 2) we can clearly see the eye—which is our target and some residue pixels. How can we find where the eye is?

The answer is its nearly circular shape. A method called the Hough transform can detect a shape in the image. The basic method detects straight lines. The generalized method can detect nearly any shape and is especially suited to the detection of a circle with a multitude of existing implementation.

Let us start from a binary image with detected points. For the sake of simplicity, we will firstly present the method for lines and show how it works. Then we will show how to use the same method for detecting circles, which is the main point of this section.

The points along the line fulfill its equations, so for each point (x_i, y_i), holds the following

$$y_i = a \cdot x_i + b, \quad i = 1, 2, ..., n, \tag{3}$$

from (3) we can calculate

$$b = (-x_i) \cdot a + y_i, \quad i = 1, 2, ..., n. \tag{4}$$

So, from a group of points on a single line in Fig. 5, we have a group of lines with a single common point. See Fig. 6. Of course, due to noise and other factors, in a typical image, there would be a group of crossings rather than a single point. A good example may be found in [7].

Generally, a type of matrix accumulator is used to calculate the crossing. Its cells span through limited parameter space with some division. Again a shape (line in a simple case) is recognized when the amount of crossings in a certain box is larger than the threshold. It is represented by the grid in Fig. 6.

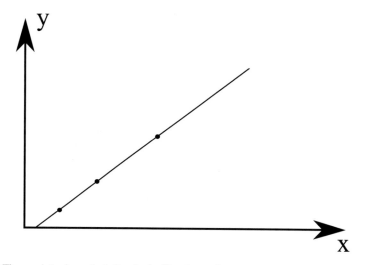

Fig. 5 Three points along single line in the Hough transform

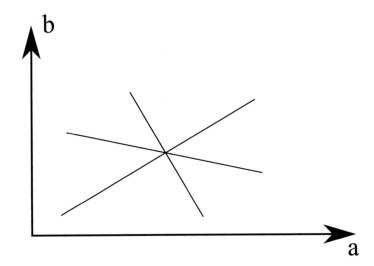

Fig. 6 Three lines crossing at a single point

The described method is simple for lines. It can be extended easily to any shape that can be parameterized. Additionally, more than two dimensions are possible – two dimensions are easy to visualize.

In case of the circle, we can use the simplest equation

$$(x - a)^2 + (y - b)^2 = r^2. \tag{5}$$

In this scenario, we require a three-dimensional accumulator and this is a little more complicated.

6 Error Analysis

In Sect. 3 we described how a thermal imaging camera works. With any measurement we expect some errors. There are many sources. Some of them, like noise, are unavoidable. Others may arise from calibration. The last type of errors come from the non-stationary behavior of the device.

The main reason is usually thermal drift. Designers try to keep it in check. In the case of the camera used it was not successful.

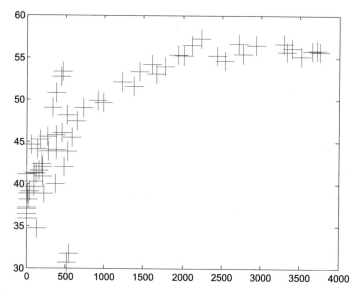

Fig. 7 Eye's measured temperature as a function of time

7 Systematic Error Removal

It is obvious that the range of expected temperatures is limited by biological boundaries. No mammal should have an internal body temperature higher than 44 °C. So a measurement of around 50 °C must be considered a gross error or some kind of systematic degradation of the measuring device. Since the time of each measurement can be established from the file's time stamp, then the measured eye's temperature can be plotted as $T(t)$. This plot is shown in Fig. 7.

Obviously, the temperature changes from one animal to another but the rise of temperature with time can be clearly seen. The shape of a possible curve looks similar to an inertial object.

The temperature of an object heated by a constant power heat source can be described by the following differential equation

$$\frac{dT}{dt} = -\alpha(T - T_q), \quad T(0) = T_0 \tag{6}$$

the solution is

$$T(t) = T_0 e^{-\alpha t} + T_q(1 - e^{-\alpha t}). \tag{7}$$

The overall shape of this solution fits the data in Fig. 7, but increasing differences between the data suggests that the error is multiplicative rather than additive in its nature.

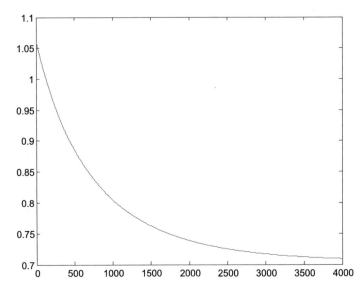

Fig. 8 Multiplicative correction coefficient

Let us consider the following correction where T_m is a measured temperature and T_c is after correction.

$$T_c = T_m \left(a - b \cdot e^{c \cdot t}\right) \tag{8}$$

a, b and c are parameters that should be fitted to the measurements. The situation is different than in typical data point fitting because deviations are useful information and the curve itself is a form of an error.

The simplest way of fitting a curve to the data are nonlinear least squares. As a result we obtain $a = 1.419$, $b = 0.477$, $c = -0.0009914$. The resulting curve can be seen in Fig. 8. We can clearly see that after the correction the range of temperatures was reduced and does not change in time. The corrected values are shown in Fig. 9.

8 Data Analysis

Measurements were taken in three different environments and in different ambient temperatures. Due to this fact, different corrections coefficients are required. The overall result can be seen in Fig. 10. One of the typical problems was the relatively small number of ill animals and the fact that temperature measurements have a tendency to cluster together, see Table 1.

We can state the hypothesis that some relationship exists between eye temperature and internal body temperature of the same animal. The biological functions of the body suggest that it would be true.

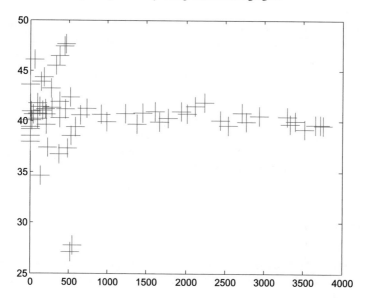

Fig. 9 Corrected data in function of time

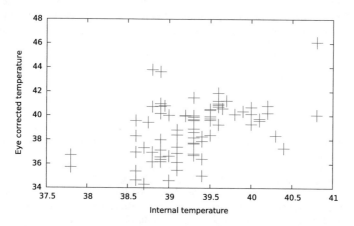

Fig. 10 After correction the pattern starts to be clearly visible

Table 1 General overview of data

Herd	Healthy	Ill	Total
A	13	3	16
B	18	17	35
C	26	4	30

The overall Pearson correlation factor is $r = 0.454$. Let us do a test for the correlation coefficient. The null hypothesis is that $r = 0$—no correlation. Then

$$t = r\sqrt{\frac{n-2}{1-r^2}} \qquad (9)$$

has t-Student distribution with $n - 2$ degrees of freedom. There were $n = 78$ measurements taken. The critical test value for significance 0.999 is 3.195. The test statistics yield $t = 4.447 > 3.195$ so we reject the null hypothesis and we accept the alternative one—the existence of correlation.

9 Conclusion

In this paper, we have shown how to use a thermal imaging camera in simple veterinarian diagnostics. We have also corrected errors that made the results unusable at the first sight.

The statistical tests have shown that the correlation is relevant. The proposed method can be used in fast temperature screening of large groups of animals. It would be possible to use a similar methodology in testing humans for the increased temperature at large-scale during epidemic events like COVID-19.

References

1. Cetinkaya, M.A., Demirutku, A.: Thermography In the assesment of equine lameness. Turk. J. Vet.Y Anim. Sci. **36**(1), 43–48 (2012)
2. Haq, A., Wilk, S., Abell, A.: Fusion of clinical data: a case study to predict the type of treatment of bone fractures. Int. J. Appl. Math. Comput. Sci. **29**(1), 51–67 (2019)
3. Hoffmann, G., Schmidt, M., Ammon, Ch., Rose-Meierhfer, S., Burfeind, O., Heuwieser, W., Berg, W.: Monitoring the body temperature of cows and calves using video recordings from an infrared thermography camera. Vet. Res. Commun. **37**, 91–99 (2013)
4. Jędruś, A., Lipiński, M.: The idea of calf's temperature monitoring (in polish). Inynieria Rol. **1**(119), 229–234 (2010)
5. Kou, H., Zhao, Y., Ren, K., Chen, X., Lu, Y., Wang, D.: Automated measurement of cattle surface temperature and its correlation with rectal temperature. PLoS ONE **12**(4), e0175377 (2017). https://doi.org/10.1371/journal.pone.0175377
6. Kowal, M., Skobel, M., Nowicki, N.: The feature selection problem in computer-assisted cytology. Int. J. Appl. Math. Comput. Sci. **28**(4), 759–770 (2018)
7. Jahne B.: Digital Image Processing. Springer, Berlin (2002)
8. Naylor, J.M., Streeter, R.M., Torgerson, P.: Factors affecting rectal temperature measurement using commonly available digital thermometers. Res. Vet. Sci. **92**, 121–123 (2012)
9. Nugent, P.W., Shaw, J.A., Pust, N.J.: Correcting for focal-plane-array temperature dependence in microbolometer infrared cameras lacking thermal stabilization. Opt. Eng. **52**(6), 61304-1–61304-8 (2013)

10. Poikalainen, V., Praks, J., Veerme, I., Kokin, E.: Infrared temperature patterns of cow's body as an indicator for health control at precision cattle farming. Agron. Res. Biosyst. Eng. Spec. Issue **1**, 187–194 (2012)
11. Polat, B., Colak, A., Cengiz, L.E., Yanmaz, H., Oral, A., Bastan, S., Kaya, S., Hayirli, A.: Sensitivity and specificity of infrared thermography in detection of subliclinical mastitis in dairy cows. J. Dairy Sci. **93**, 3525–3532 (2010)
12. Rafajowicz, E., Rafajowicz, W.: Introduction into industrial image processing (in Polish). Oficyna Wydawnicza Politechniki WrocławskiejWrocław (2010)
13. Rząsa, A., Jawor, P., Rafajłowicz, W., Tybor, K., Trzaskacz, M.: Thermography as a tool in veterinary (in polish). Weterynaria w Terenie **3**, 80–83 (2015)
14. Schaeffer, A.L., Cook, N.J., Chuch, J.S., Basarab, J., Perry, B., Miller, C., Tong, A.K.W.: The use of infrared thermography as an eary indicator of bovine respiratory disease complex in calves. Res. Vet. Sci. **83**, 376–384 (2007)
15. Rząsa, A., Jawor, P., Pieczewska, B., Rafajłowicz, W.: Infrared thermography as a tool in veterinary practice. XV Kongres Polskiego Towarzystwa Nauk Weterynaryjnych, Lublin. 22-24.09.2016, p. 247 (2016)

Intruder Detection on Mobile Phones Using Keystroke Dynamic and Application Usage Patterns

Michał Szczepanik◉ and Ireneusz Jóźwiak◉

Abstract Security through biometric keystroke and user action analysis on mobile phones is a new approach. Most of the existing solutions track only GPS location, which allows detecting theft. None of them support spy detection or short access of the intruder when the phone is away from its owner. In this paper, Authors present a solution that allows detecting intruder via analysis of typical user's actions in the applications and way of writing - keystroke dynamics. Presented solutions can be run as an application on Android mobile devices and currently is distributed via Google Play Store alpha channel for testing by the limited number of users.

Keywords Mobile security · Keystroke dynamics · Intruder detection · Biometrics

1 Introduction

The number of smartphone users keeps increasing every day for the past few years, and this is the expected trend for also next years. As per comScore's 2017 report [1], the average adult spends over 3 h on the smartphone every day. It is used not only for entertainment but also for work, social life, and financial management. This is proved by research and statistic collected by AdMob team [2], according to them users use a mobile browser to:

- Connecting with friends and social media—45%,
- Downloading files—46%,
- Visiting websites—9%.

M. Szczepanik (✉) · I. Jóźwiak
Faculty of Computer Science and Management,
Wroclaw University of Science and Technology, Wroclaw, Poland
e-mail: michal.szczepanik@pwr.edu.pl

I. Jóźwiak
e-mail: ireneusz.jozwiak@pwr.edu.pl

© The Author(s), under exclusive license to Springer Nature Switzerland AG 2021
J. Korbicz et al. (eds.), *Advances in Diagnostics of Processes and Systems*,
Studies in Systems, Decision and Control 313,
https://doi.org/10.1007/978-3-030-58964-6_11

Additional research presented in the FICO report [3] proves that:

- 20.4% of users use mobile banking,
- 58% would like to start using it, but they are not currently doing so,
- 1/3 users in the 16–35 age group currently use these services.

These numbers increase almost every year, additionally, solutions for payment, like Google Pay are very popular nowadays.

Presented above researches confirm that uses usually store much more private data on the phone than on any other device. The level of trust is also psychologically high because most of the users store phone near to their heart [4]. According to that a lot of intruders select mobile phones as a target to collect private or sensitive data. Access is possible by spy software or malware, but nowadays most cybersecurity and Anti-virus providers focus also on security on mobile phones and allow to detect this kind of software [4]. Physical access to the phone is easy and allow us to spy much more data than applications because it is not limited by system permissions.

1.1 The Problem

Authorization on smartphones is usually done by pin (usually 4 numerical digits), password (alphanumeric), fingerprint recognition or quite popular in new phones face recognition. In the case of fingerprint or face recognition the alternative solution exists, which is password or even pin as the biometric analysis is not always able to correctly recognize the user. This scenario can be easily forced even by an intruder as:

- Fingerprint recognition system cannot recognize wet or dirty fingerprints [5],
- Face recognition system does not work in dark places or when the camera caught more than two faces for most common used algorithms.

Described above edge cases scenarios are rather typical during daily basic usage which means user need to use pin or password instead of biometric methods.

Digits needed to unlock the phone can be easily caught by potential intruders only by observation. Knowledge of password allows him/her to use it later when the user leaves the phone for example on a desk without protection.

1.2 The Goal

Detection of the attack described in Chap. 1.1 is an important problem to solve and the main research of this paper. Potential intruder uses a phone differently as he/her first open applications, which store a lot of data about the user, in a short period. The main targets are applications for email, banking, social media, messaging and even photo gallery.

Additionally, usage of keyboard is one of biometric characteristic which can allow recognizing user, and it is quite popular for physical keyboard [6]. This type of solution was ported to virtual keyboards, but current devices have a different type of them like a normal, swipe or autocomplete. As a solution for intruder detection Authors propose two mechanisms:

- Keystroke recognition for the virtual keyboard,
- Application usage analysis.

An additional goal is to implement this solution for Android system and release the practical application on the market, so all limitations related to the operating system and features need to be taken into account.

2 Existing Solutions

One of the best-known existing solutions for detecting the thief of the phone is iGuard. It uses a motion segmentation algorithm [7] and Markov Chain based model to track the behavior of a smartphone user. According to this model, iGuard instantly alarms once the tracked data deviate from the smartphone owner's habit. It allows for detecting for example if someone else takes the phone from pocket or bag. Unfortunately, this system does not allow us to detect if someone else uses the phone when the user is away from it. An additional disadvantage of the algorithm is battery usage which increased during movement detection and depends on the device. Based on iGuard team battery usage should increase only 2% on Samsung S4, but the usage of this app on Motorola G6 Plus shows that battery consumption of iGuard is 5–10% and depends on movement.

The next solution is Prey Anti Theft [8] which works similarly as Google Find my Phone. This location-based solution has some standard features to track the phone like:

- Control Zones: Create areas on a map that alert device movement in and out of them,
- GPS Geolocation: Pin-point accuracy reflected on a map, with GPS coordinates,
- Location History: Verify device movements in time and detect suspicious movements.

This solution allows users to detect where his phone is, but the application only works when there is access to the internet, GPS and cellular network. A good additional feature of the app is the option to use cameras to make photos periodically, which may help to identify the intruder.

Another specialized application to protect phone and data is Cerberus anti-theft [9]. It includes features like the location history, tracking your device, lock it with code, starts alarm with a message, displays a message, calls it, getting SMS and Call logs, and wiping all user's data.

It also makes a photo of a potential intruder when the user is wrongly authenticated by the lock method.

There are many other applications that allow to track the phone or provides similar anti-theft solutions [10]. None of them check the user while using the phone. Additionally, most of them require internet access to do the action and block or wipe the phone (remove all sensitive data). This requirement is a huge limitation for example when the user is abroad (high cost of Internet access), has no active data plan or theft disconnects phone faster than any action was done.

According to that analysis can be seen that there is no on-device solution that will block the phone when it is under unknown user control.

3 Keystroke Dynamics

First of the proposed solution based on keystroke recognition which is a typical solution for hardware keyboard [11, 12]. Studies have shown that each person has certain unique features that can be manifested in a variety of typing rhythms [13–15]. The delay between pressing each key or the duration of holding individual keys while typing may distinguish the user from the others. This is behavioral biometry that can be used in many ways, such as identifying or even detecting emotions. So far, many implementations have been created focusing on the analysis of typing dynamics. One of online keystroke recognition solution was presented in [16]. It sends delays between pressing keys on the keyboard additionally with a password. Collected data are compared with pattern and if the differences between delays are too big the access is not granted. This type of algorithm can be also used for virtual keyboards since it only uses delays for this detection. Shen et al. [17] proposed a solution based on Dynamic Keystroke Pressure-Based which allows continual identity verification during the whole time of writing. Trojahn [18] proposed solutions for the full QUERTY virtual keyboard, which based on press delays and other features that can be collected from the display sensors.

Current mobile devices provide not only data about which key (button) was pressed, but additionally coordinates of touch, size of the touched area (size of finger), and press strength. This kind of data allows implementing similar solutions like exist for hardware keyboard. Of course, data should be collected separately for both device orientations (portrait and landscape) as the way of usage is different. For example, the portrait mode is usually used for one-hand operation and for writing in landscape orientation users usually use two hands.

3.1 Method Based on Basic Features (BFA)

Standard keystroke dynamic analysis based on two events like key pressing time and time between pressing two different digits. In basic analysis, Authors decide to

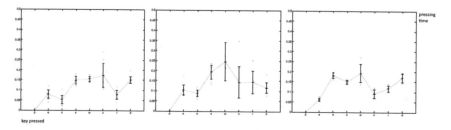

Fig. 1 Delays between keystrokes when entering the word password by three different people

Fig. 2 A data received from the virtual keyboard

only use time between pressing keys corresponding to alphanumeric characters as any other side effects such as the length of the key being held by the writer or the dynamics of a particular key are not possible here to measure due to the too-small screen of the device and minimalist icons corresponding to individual keys, whose pressing is associated only with a brief touch. To better visualize the proposed solution, the Fig. 1 showing the delay between each key press and its duration. The X-axis represents the individual letters of the password, the Y-axis times in milliseconds.

The enrolment process has four different steps, like for any other biometric recognition technique:

- Data acquisition + enrolment samples,
- Preprocessing,
- Feature extraction,
- Storage.

The data collected form the device are presented on Fig. 2.

During few attempts to obtain the best data sample, it was noticed that within such an algorithm of data collection there are time anomalies. Let us assume the case in which the writer stops for a moment and do not press any keys. For that case special limit was included which is 1500ms. If the delay between two keypresses is bigger than then whole this data vector is ignored.

The user model is the model with which the thief model will be constantly compared to determine the result of the final algorithm. The input argument for these calculations is the aforementioned keymap (character pairs) and values (time lists). The output parameter is the same keymap, while the values are calculated time anomalies, called anomaly points. In order to obtain anomalies for all character pairs, the Euclid algorithm was used.

The Euclidean distance algorithm creates a p-dimensional space, where p is the number of times received for each pair of characters. Each sample is a point in the created space, therefore training data is modeled as a point cloud, where the average vector from this data is defined as the cloud center. In this phase, Euclidean distances (1) between test vectors and average vectors obtained from the same data are calculated.

$$A = \sqrt{\sum_{i=1}^{p}(x_i + y_i)}, \tag{1}$$

where:

p—is number of times recorded for each pair of characters,
x_i—is i-th feature of the vector of average values,
y_i—is i-th feature of the test vector.

The pattern recognition model calculates the anomaly result for each sample and creates a threshold from the anomaly results obtained for the owner model. To create the threshold, all previously calculated anomalies for each character pair are added and divided by the number of anomalies.

The classification phase consists of assessing the newly received test sample. Anomalies are calculated for the potential thief's model. If the thief's threshold is greater than the phone owner's threshold, the function that calculates the probability of the device being intercepted is true. This means that the data sample was written by someone other than the phone owner who generated the correct data model. Otherwise, the algorithm proceeds to calculations related to the next series of data. This mechanism has been designed to learn with the user. The owner of the phone can constantly use it normally while the background application is constantly learning and updating its model. This mechanism is useful because the user's typing pattern may change over time.

3.2 Method Base on Additional Features (AFA)

The first method can be used on all smartphones as it does not require any additional features (touch area or pressure), which are not available for most of the old devices. New devices allow developers to get additional data about user interaction like touch strength or size of the touched area, both these features can be used to verify user during writing or phone usage. This method provides a modification to the previous one and stores additional data which are data regarding touches, like strength (5 level scale) and average radius of the touch area.

The size of the area allows checking if the person which use the phone has similar fingers and way of touching the screen. Basic tests show that the difference of size registered by the device can have 3% differences and this threshold level was selected in the algorithm.

The strength of touch (pressure) is much more problematic as the value of it can be different in many cases for example it depends on our current emotions and way how the phone is held. The value which can be measured is the differences between touch strength in different areas of the screen as this is quite similar and does not depend on current emotions. According to that knowledge the screen was spat to the matrix of areas with size 16×9 (in final implementation it depends on aspect ratio). For each area, the average touch pressure is measured, and it is used to measure the difference and correlation between areas. This correlation is used additionally in user verification during his/her phone usage.

3.3 Method for Swipe Like Keyboard (SKA)

The main problem of keystroke dynamics during normal phone usage is the usage of swipe like keyboards. Using them users instead of pressing buttons related to litters use swipe gestures instead. Unfortunately, this kind of keyboard previously presented solution will not work and data representation needs to be different.

Representation of each letter can be done by coordinates (x, y) the characters and coordinates, which need to be stored, are defined by changes in swipe direction. This algorithm will work in a similar way how to swipe keyboard detector works. The time of swipe between two points represents the same correlation as was used in the first algorithm. The only difference is data representation which is related to exact coordination. The biggest problem of this algorithm is analysis required by mapping points to letters which mostly depends on keyboard and device.

3.4 Methods for Autocomplete (+AC)

Autocomplete is a very popular feature of mobile device keyboards which makes the writing process much faster. Unfortunately, analysis of autocomplete usage needs data related to specific words, according to Google requirements and regulations this kind of statistic cannot be store by applications and this solution cannot be fully implemented. The only data which can be stored by the applications are general statistics of the usage of these futures, as the average length of words when this feature is used.

4 Application Usage Analysis (AUA)

A potential intruder can not only use the keyboard but mostly he/she will focus on the apps in which important data are stored. Each user has a specific path in the app which he/she is doing just after opening it. For example, in email applications usually, we check recent messages and do not scroll a lot, which can be typical behavior of the intruder. For banking applications, user rather will do small transfer or only check the amount of the money on the account than do the big transfer. Each app has a spe-cific typical path of action which user do on a daily basis and any other way of usage can be detected as an anomaly.

Each user action is registered as one of gesture: click or swipe in the area or many areas (in case of swipe) of the screen. For an implementation aspect ratio of the screen (16×9 or 16×10 $16 \times 12(4 \times 3)$) define the matrix of areas used by the algorithm.

To design basic models for TensorFlow Lite [19], which is a deep learning framework for on-device inference, data from 20 users were collected. Data include two weeks usage of 10 most popular applications for email (Gmail, Outlook), messaging (Facebook messenger, Whatsapp, Messages, Hangouts), financial (mBank, Revolut), photo browser (Photos) and home automation (Home). Each of these applications is from a specific category and most of the other applications from the same category are used in a similar way, for example, email clients, messaging applications.

These ready models are used on the device to additionally learn and detect specific for its user behaviors. Data regarding which model should be used can be defined based on Play Store data and application category in it. The current list of available models is limited to email clients, messaging, financial, photo browser and home automation applications as all of them store most sensitive and private data about the user which are the target of the intruder.

5   Proposed Application

The final solution combines these two approaches as they are cover two main areas in which intruder can be detected before he/she will copy or get knowledge about sensitive or private data about the user. Keystroke dynamic recognition is used on the password screen and during writing a text. The detailed algorithm used in this step depends on the device (features that can be collected) and keyboard type (swipe sup-port). For passwords, almost all keyboards block swiping, so verification is easy and does not require detailed knowledge about buttons areas on the virtual keyboard. In case when an intruder gets the phone, which is unlocked or unlocked it in other way algorithm related to application usage is run. It monitors only apps selected by the user to be protected to reduce battery consumption and usage of phone resources.

6   Experiment

Each authentication based on biometric has the disadvantage that some persons are falsely rejected (False Reject Rate - FRR) and others are false accepted (False Accept Rate - FAR). This means that people who are truly a user of the system can be rejected and some intruders can be accepted. Both error rates should be as low as possible. The problem is that both cannot be zero at the same time and they have to be balanced for the special scenario.

In some literature, the EER (Equal Error Rate) is used instead of FAR and FRR. EER is the point where FAR and FRR are equal. But the challenge is to find the right threshold for a situation and use case. Some systems need a low FAR (high threshold) that no intruder can access the system. Other systems have to have high usability, so it is important that a user does not need to authenticate him five times. Additionally, both system characteristics can be easily measured and calculated, see (2) and (3).

$$FAR = \frac{number\ of\ false\ acceptances}{number\ of\ impostoridentification\ attempts} \qquad (2)$$

$$FRR = \frac{number\ of\ false\ rejections}{number\ of\ impostoridentification\ attempts} \qquad (3)$$

For experiment four different type of devices were used:

– Pixel 2 XL with Android 9.0
– Pixel 3 with Android 10.0
– Motorola G6 Plus with Android 9.0
– Asus ROG Phone II with Android 9.0.

Tests were done on 10 different users during the one-month period. Algorithms were selected based on available features of the screen and keyboards. For experiment different combinations of algorithms were used:

– BFA—basic feature algorithm, described in Sect. 3.1
– AFA—additional features algorithm, described in Sect. 3.2
– SKA—swipe keyboard algorithm, described in Sect. 3.3
– +AC—autocomplete support included in the main algorithm, described in Sect. 3.4
– AUA—application usage analysis, described in Sect. 4.

Test done on Pixel 2 XL are presented in Table 1. This phone does not support any additional features about touches, and it is limited to use only BFA solution. Additionally, the big size of the screen caused anomalies in AUA solution and the value of FRR is big.

For Pixel 3 (see Table 2) all algorithms provide very good results. The support for autocomplete (+AC) is also acceptable but this mechanism generates a lot of shorts strings of characters provided by the user and sometimes for not long text data collected from keyboard need to be ignored as number of characters is less than 5.

On Motorola (see Table 3) similar situation as on Pixel 2 XL occurs for applications usage analysis as screen is also big.

Asus ROG is gaming smartphone and provide access to a lot of precise data related to touches on the screen like size, pressure. This allow to use them in AFA solution and make it the best solution for intruder detection. Even not standard aspect ratio of the phone 19.5:9 was not big problem for the algorithms Table 4.

Table 1 Tests done on Pixel 2 XL (*Source* Own work)

	FAR(%)	FRR(%)
BFA	1,25	2,25
BFA+AC	1,75	2,50
SKA	2,50	5,25
AUA	2,25	10,25
AUA+BFA+AC	1,25	2,00

Table 2 Tests done on Pixel 3 (*Source* Own work)

	FAR(%)	FRR(%)
BFA	1,75	2,0
AFA	0,75	1,50
AFA+AC	0,75	2,00
SKA	2,50	4,75
AUA	2,00	4,25
AUA+AFA+AC	0,75	1,00

Table 3 Tests done on Motorola G6 Plus (*Source* Own work)

	FAR(%)	FRR(%)
BFA	1,25	2,00
AFA	1,00	1,50
AFA+AC	1,25	2,25
SKA	2,20	4,50
AUA	2,50	9,50
AUA+AFA+AC	1,00	1,00

Table 4 Tests done on Asus ROG phone II (*Source* Own work)

	FAR(%)	FRR(%)
BFA	1,50	2,25
AFA	0,50	1,25
AFA+AC	0,75	1,25
SKA	2,50	4,75
AUA	2,00	5,25
AUA+AFA+AC	0,75	1,00

Presented data confirms that proposed solutions are promising and may allow to detect unauthorized access to the smartphone.

7 Conclusion and Future Work

Proposed algorithms are an innovative way of intruder detection for mobile devices. According to market research, there is no other solutions which allow detection of physical spy attack on the mobile phone. It uses two approaches: keystroke dynamics and application usage analysis, which allow detecting intruder during phone usage.

Currently, the application is available in the alpha channel of Google Play Store [20] for a limited number of users - currently 52 (2019/12/30). First feedback is very good for young and middle-age users, additionally, values of FAR and FAR are quite similar to values from the experiment. Only in a group of users with age 60+ a lot of sessions with false rejection was detected. Expected reasons for that issue are related to the speed of writing, as delays between key pressing are big for this group.

As future work support for tablets is planned and research regarding auto parental control as algorithms may detect users when the phone is shared, for example with children.

References

1. comScore.: https://www.comscore.com/. Last accessed 12 Dec 2019
2. AdMob serves a billion.: http://www.mobileent.biz. Last accessed 15 Dec 2019
3. Lysik, L., Machura, P.: The role and importance of mobile technology in everyday life of the 21st century (in Polish). http://www.mediaispoleczenstwo.ath.bielsko.pl/art/04_lysik_machura.pdf. Last accessed 30 Dec 2019
4. Kedziora, M., Gawin, P., Szczepanik, M., Jóźwiak, I.: Malware detection using machine learning algorithms and reverse engineering of Android Java Code. (2019). https://doi.org/10.5121/ijnsa.2019.11101
5. Szczepanik, M., Jóźwiak, I.: Fingerprint recognition based on minutes groups using directing attention algorithms. In: Rutkowski, L., Korytkowski, M., Scherer, R., Tadeusiewicz, R., Zadeh, L.A., Zurada, J.M. (eds) Artificial Intelligence and Soft Computing. ICAISC 2012. Lecture Notes in Computer Science, vol. 7268. Springer, Berlin, Heidelberg (2012)
6. Monrose, F., Rubin, A.D.: Keystroke dynamics as a biometric for authentication. Future Gener. Comp. Syst. **16**, 351–359 (2000)
7. Jin, M., He, Y., Fang, D., Chen, X., Meng, X., Xing, T.: iGuard: A real-time anti-theft system for smartphones. In: IEEE INFOCOM 2017—IEEE Conference on Computer Communications, Atlanta, GA, pp. 1-9 (2017). https://doi.org/10.1109/INFOCOM.2017.8057021
8. Prey Project.: https://preyproject.com/. Last accessed 18 Dec 2019
9. Cerberus.: https://www.cerberusapp.com/. Last accessed 18 Dec 2019
10. Szczepanik, M., Jóźwiak, I.J., Jamka, T., Stasiński, K.: Security lock system for mobile devices based on fingerprint recognition algorithm. In: Świątek, J., Borzemski, L., Grzech, A., Wilimowska, Z. (eds) Proceedings of 36th International Conference on Information Systems Architecture and Technology—ISAT 2015—Part III. Advances in Intelligent Systems and Computing, vol. 431. Springer, Cham (2016)

11. Deng, Y., Zhong, Y.: Keystroke dynamics advances for mobile devices using deep neural network in Recent Advances in User Authentication Using Keystroke Dynamics Biometrics, pp. 59–70 (2015)
12. Maiorana, E., Campisi, P., González-Carballo, N., Neri, A.: Keystroke dynamics authentication for mobile phones. In: Proceedings of the ACM Symposium on Applied Computing, vol. 21–26 (2011). https://doi.org/10.1145/1982185.1982190
13. Lee, H., Hwang, J.Y., Kim, D.I., Lee, S., Lee, S.H., Shin, J.S.: Understanding Keystroke Dynamics for Smartphone Users Authentication and Key-stroke Dynamics on Smartphones Built-In Motion Sensors. Secur. Commun. Netw. **2018**(2567463), 10 (2018)
14. Alshanketi, F., Traoré, I., Awad, A.: Multimodal mobile keystroke dynamics biometrics combining fixed and variable passwords. Secur. Privacy **2**, e48 (2019)
15. Shahzad, F.: Low-cost intruder detection and alert system using mobile phone proximity sensor. In: 2017 International Conference on Innovations in Electrical Engineering and Computational Technologies (ICIEECT), pp. 1–5. Karachi (2017)
16. Jiang, C.H.: Keystroke statistical learning model for web authentication. http://citeseerx.ist.psu.edu/viewdoc/download?doi=10.1.1.314.1573&rep=rep1&type=pdf. Last accessed 30 Dec 2019
17. Shen, S., Lin, S., Kang, T., Chien, W.: Enhanced keystroke dynamics authentication utilizing pressure detection. In: 2016 International Conference on Applied System In-novation (ICASI), pp. 1–4. Okinawa (2016). https://doi.org/10.1109/ICASI.2016.7539947
18. Trojahn, M.: Biometric authentication through a virtual keyboard for smartphones. Int. J. Comput. Sci. Inf. Technol. **4**, 1–12 (2012)
19. TensorFlow Lite.: https://www.tensorflow.org/lite. Last accessed 12 Dec 2019
20. Spydetector.: https://play.google.com/store/apps/details?id=com.apartapps.spydetector. Last accessed 30 Dec 2019 (restricted for alpha channel users)

Medical Applications

Application of Deep Learning to Seizure Classification

Krzysztof Patan and Grzegorz Rutkowski

Abstract The paper deals with the application of deep neural networks to design a computer-aided system capable to detect epileptic seizures. A deep long short-term memory (LSTM) is used to discover dependencies between samples of processed electroencephalogram (EEG) signal at different time instances. Two classification schemes are investigated and compared: sequence-to-sequence and sequence-to-label classifiers. The research was carried out using real EEG recordings of epileptic patients as well as healthy subjects prepared with the cooperation of the medical staff of the Clinical Ward of Neurology of the University Hospital of Zielona Góra, Poland.

Keywords EEG signals · Seizure detection · Deep learning · Long short-term memory · Classification

1 Introduction

Epilepsy is a neurological disorder which is more and more apparent in the present world. It is estimated that approximately 1–2% of world's population has epilepsy, about 5% of people may have at least one seizure during their lifetime, and about 25% of epileptic patients cannot be treated efficiently by any available therapy [2]. Epileptic seizures can be divided into partial, generalized, unilateral and unclassified [20]. In turn EEG recordings of epileptic patients can be classified into two categories: *inter-ictal* representing an abnormal activity recorded between epileptic seizures, which has the form of occasional short-term transient waves, and *ictal* showing an activity recorded during the seizure in the form of long-term polymorphic waves [20]. To date, many different seizure detection systems have been proposed [1, 3, 5, 14]. The widely used method of EEG analysis in the time-domain is visual inspection.

K. Patan (✉) · G. Rutkowski
University of Zielona Góra Institute of Control and Computation Engineering,
ul. Szafrana 2, 65-516 Zielona Góra, Poland
e-mail: k.patan@issi.uz.zgora.pl

© The Author(s), under exclusive license to Springer Nature Switzerland AG 2021
J. Korbicz et al. (eds.), *Advances in Diagnostics of Processes and Systems*,
Studies in Systems, Decision and Control 313,
https://doi.org/10.1007/978-3-030-58964-6_12

Unfortunately, the length of the EEG record can reach one hour then a neurologist can spent a lot of time observing and analysing the record. What is even more noteworthy, even diagnoses carried out by the same expert but in different periods of time may differ significantly. Systems working in the time-domain can investigate statistic properties of EEG signals as basic or regularity statistics [23], the approximate entropy [3], pattern match regularity statistics [17], and others. Unfortunately, a majority of papers deal with long-term seizures. On the other hand, a large number of publications is devoted to the analysis of EEG recordings in the frequency or time-frequency domain. Very popular methods used in this scope are Discrete Wavelet Transform [6, 19, 22], Matching Pursuit [1, 23] and recently a Stockwell Transform [12, 14, 15]. Unfortunately, methods working on the time-frequency representation are time-consuming. Then it is a need to develop novel methods characterized by a fast processing time. Recently, the popular approach to solve classification problem has become deep learning. Deep learning methods, especially different variants of convolutional neural networks (CNN) proved their usefulness in many different classification tasks [8, 9, 11]. However, CNN are rather used to classify two-dimensional images then they cannot be used to process time-sequences directly. Hopefully, to process time-sequences another deep learning method called deep long short-term memory can be applied with success [13, 16, 21].

Our objective is to detect the inter-ictal seizures pending less than 1 s. Due to the fact that considered seizures are characterized by a short-time occurrence the problem of seizure detection becomes non trivial. The main goal of this work is to design an automated computer system for diagnosis of epileptic seizures. Such a computer-aided diagnostic system renders it possible to reduce the time spent by an expert to find the characteristic graphoelements representing epileptic seizures. The present work is the extension of the ideas presented in the previous work of the authors [13]. To solve the problem it is proposed to apply LSTM networks to develop a seizure detection system. Two ideas are investigated: a sequence-to-sequence mapping and a sequence-to-label classification. All experiments are carried out using MATLAB© software. The contribution of the paper can be listed as follows.

1. Application of deep LSTM to seizure detection using two detection schemes: a sequence-to-sequence and a sequence-to-label classification,
2. The selection of the optimally performed neural network structure,
3. Thorough evaluation of both investigated classification schemes as well as comparative studies.

2 Epilepsy Detection

The purpose of the detection system is to point out whether a seizure occurs or not. Moreover, the important issue is to provide the occurrence time of a seizure. The input data was the EEG record containing 16 channels. It means that the input space consists of 16 time sequences. In this work we deal with short-time seizures, usually

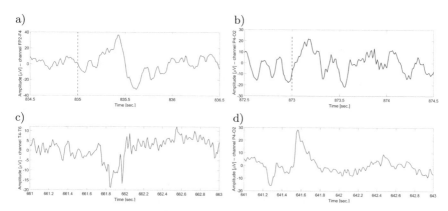

Fig. 1 EEG examples: seizures (**a**, **b**) and normal cases (**c**, **d**)

of the length shorter that one second. In Fig. 1a and b we can see the examples of short-time seizures pointed out by the expert. The beginning of the seizure is marked by red–dashed line. Clearly, we can observe the change in the amplitude as well as in other signal statistics. However, in Fig. 1c and d there are shown examples of the normal course of the EEG signal. In these two cases we can also observe changes in the amplitude and statistical characteristics of the signal, however this time there is no seizure at all. The presented illustrative examples comparison clearly show that the problem of detecting short-time seizures is very difficult to solve and efficient detection methods are highly desirable. What is even more noteworthy, after consultations with specialists we found out that the computer application should be developed in order to minimize the time spent by a specialist to analyze EEG records. For example, the visual inspection of one EEG record is time-consuming and can last a dozen of minutes. Therefore, taking into account the following facts:

– A neurologist provided the estimated time of seizure occurrence only. Then, the real seizure can be observed a little bit earlier or a little bit later;
– A neurologist did not provide the end time of seizures. Then, the correct length of a seizure is not known;

we assumed that the length of a seizure is equal to one second. This assumption can facilitate the analysis as well as the implementation of a detection system. Summarizing, the analysis is performed using a moving window extracting the portion of data for analysis. To each extracted sequence a suitable label needs to be assigned. In this paper we consider two possible solutions:

– Sequence-to-sequence mapping—to every input time-sequence o the length n it is assigned an output time-sequence also of the length n (Fig. 2a). Each value of the output sequence determines a state of the patient at a given time-instance,
– Sequence-to-label mapping—to every input time-sequence of the length n it is assigned a output label determining the patient state (Fig. 2b).

The patient state may be marked as either 'epileptic' or 'normal'.

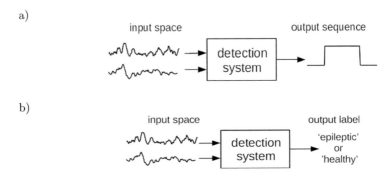

Fig. 2 Idea of sequence-to-sequence (**a**) and sequence-to-label (**b**) classification

To evaluate the quality of the detection systems the following indexes are used:

– A sensitivity (a true positive rate) for measuring the rate of correctly identified seizures:

$$tpr = \frac{n_{tp}}{n_p} 100\%, \tag{1}$$

where n_{tp} represents the number of detected seizures, n_p is the total number of seizures;

– A specificity (a true negative rate) for measuring the rate of correctly detected healthy cases:

$$tnr = \frac{n_{tn}}{n_n} 100\%, \tag{2}$$

where n_{tn} stands for the number of sequences classified as normal ones, and n_n is the total number of healthy cases;

– A total accuracy:

$$acc = \frac{n_{tp} + n_{tn}}{n_p + n_n} 100\%. \tag{3}$$

In cooperation with the medical staff of the Clinical Ward of Neurology of the University Hospital of Zielona Góra, Poland, a database of neurological disorders was prepared. The data were acquired from both epileptic patients and healthy subjects. With the help of neurologists, 588 seizures from 104 patients, both females and males, were recorded and analyzed. Simultaneously, the database includes recordings acquired form 71 healthy subjects. Each EEG record consists of 16 signals.

3 Deep Neural Networks

In order to process time-sequences it is recommended to use a neural network with dynamic properties able to catch time dependencies between values of the subsequent time instances. Among many recurrent neural networks proposed to date, a long short-term memory [4, 13] is a very interesting and desirable neural model as it can cope with the problem of the blowing-up or vanishing gradient. A fundamental LSTM architecture includes: a memory cell and three gates namely input, output and forget ones. The memory cell is responsible for storing data for arbitrary time periods. In turn, gates control data flow through connections of the whole model. The basic representation of the single LSTM unit is given by a set of equations. The input gate is represented as:

$$i(k) = \sigma_s(W_i x(k) + V_i h(k-1) + b_i), \tag{4}$$

where W_i, V_i represent weight matrices and b_i is the bias vector of the input gate, $x(k)$ and $h(k)$ are the input and the output of the LSTM unit, and σ_s stands for the activation function. In a similar fashion, the output gate

$$o(k) = \sigma_s(W_o x(k) + V_o h(k-1) + b_o) \tag{5}$$

and the forget gate

$$f(k) = \sigma_s(W_f x(k) + V_f h(k-1) + b_f) \tag{6}$$

are described, where W_o, V_o and b_o are parameters of the output gate and W_f, V_f and b_f are parameters of the forget gate. Using (4), (5) and (6), the memory cell is given by

$$c(k) = f \circ c(k-1) + i(k) \circ \sigma_c(W_c x(k) + V_c h(k-1) + b_c) \tag{7}$$

and the cell output:

$$h(k) = o(k) \circ \sigma_h(c(k)), \tag{8}$$

where W_c, V_c and b_c are weight matrices and bias vector of the memory cell, σ_c and σ_h stand for the activation functions, $\circ$ represents the element-wise product, and the initial values are $c(0) = 0$ and $h(0) = 0$.

Recently deep learning has become a popular and effective paradigm for neural network training especially when the network size is large [10]. In this framework a deep LSTM network is one of the most frequently used at the moment [16]. A deep LSTM consists of one or more LSTM layers, a fully connected layer, a softmax layer and an output layer. A scheme of the detection system based on deep LSTM is presented in Fig. 3. The input sequences are processed to the LSTM layer, where LSTM can be used to remember long-short term dependencies between samples in

Fig. 3 A block scheme of
the classification system

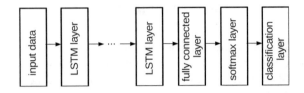

the input data. This knowledge flows through a fully connected layer to a softmax
layer. The softmax layer is used to assign decimal probabilities to each class of
the considered problem. Then the layer has the same number of units as the output
layer. Furthermore, these decimal probabilities have to add up to 1. The operation
performed by this layer can be represented as follows:

$$P(y = j|x) = \frac{e^{x_j}}{\sum_{k=1}^{K} e^{x_k}}, \quad j = 1, \ldots, K, \tag{9}$$

where j is the class index, K is the number of classes, x is the softmax layer input.
Finally, a classification layer takes the values from the softmax layer and assigns each
input to one of the K mutually exclusive classes using the cross-entropy function:

$$loss = -\sum_{i=1}^{N} \sum_{j=1}^{K} t_{ij} \ln(y_{ij}), \tag{10}$$

where N stands for the number of samples, t_{ij} is the indicator that the ith sample
belongs to the jth class, and y_{ij} is the output of the softmax unit.

In order to improve generalization abilities of the model, a dropout method can
be applied [18]. In the framework of deep LSTM a dropout layer can be placed after
each LSTM layer. Then, at each training iteration a number of processing units is
either dropout of the network with the probability $1 - p$ or kept with the probability
p. The only parameter needed to be set is the probability p.

4 Sequence-to-Sequence Classification

4.1 Model Developing

To train the neural network the available data set containing 1176 EEG records
including 588 epileptic and 588 healthy cases was split into the training and testing
sets. Both sets were separable one from another. Each investigated neural classifier
was trained 10 times using the repeated hold-out method with the dividing ratio
equal to 0.5. A crucial problem when dealing with neural network training is to
properly set the neural network topology. The input space is equal to the number of

Table 1 Sequence-to-sequence: investigated structures

Network	Number of nodes				
	1st layer	2nd layer	3rd layer	4th layer	5th layer
net1	40	–	–	–	–
net2	60	–	–	–	–
net3	100	–	–	–	–
net4	400	–	–	–	–
net5	60	50	–	–	–
net6	100	70	–	–	–
net7	50	30	10	–	–
net8	50	50	50	–	–
net9	100	70	30	–	–
net10	30	30	30	30	–
net11	60	50	40	20	–
net12	70	70	70	70	–
net13	100	70	50	30	–
net14	50	50	50	50	50
net15	100	80	60	40	20

recorded time-sequences. In the considered case the number of measured channels is 16. As mentioned earlier, in a typical situation the seizure time is shorter than one second. Then we used training sequences of the length of 3 s because in such a setting we can indicate the beginning as well as the length of a seizure. In Fig. 2a we can see the exemplary profile of the desired output sequence. During the experiment we investigated different neural network settings. A specification of investigated models is presented in Table 1. For all structures σ_c was selected as the hyperbolic tangent function, σ_s was the sigmoid function, and σ_h was the linear one. Each neural network was trained using the ADAM algorithm [7] with the initial learning rate set to $\eta = 0.02$. After training, each model was properly evaluated. Taking into account the specific form of the output representation as well as its imprecise nature we proposed the following criteria helpful in sequence classification:

Criterion 1: A sequence is classified as "epileptic" if both indexes tpr and tnr achieve the value greater than 50%,

Criterion 2: A sequence is classified as "healthy" if the index tnr achieves the value greater than 50%,

Criterion 3: A sequence is properly classified if the index acc achieves the value grater or equal to 85%.

Model evaluation is shown in Table 2 where the best results are marked by the bold face. As each network was trained 10 times for each criterion we derived two marks: the average classifier quality calculated over all trials and the best achieved result. Taking into account **Criterion 1** it is clear that deeper networks outperform simpler

Table 2 Evaluation results

Model	Criterion 1		Criterion 2		Criterion 3	
	Best [%]	Average [%]	Best [%]	Average [%]	Best [%]	Average [%]
net 1	64.8	64.8	81.5	81.1	72.9	70.9
net 2	69.4	67.8	**81.9**	81.4	77.6	76.7
net 3	66.0	64.9	**81.9**	**81.7**	77.3	76.5
net 4	73.6	64.7	**81.9**	80.9	79.1	71.6
net 5	88.1	81.8	80.6	79.7	86.4	82.9
net 6	71.6	70.7	**81.9**	81.3	80.4	79.3
net 7	95.3	91.3	79.8	78.1	92.4	88.9
net 8	95.2	91.4	79.2	78.0	91.2	87.9
net 9	96.6	92.3	79.1	77.9	94.6	89.3
net 10	98.3	**97.9**	78.0	75.5	95.2	94.3
net 11	98.8	96.4	79.2	77.1	**96.0**	**94.8**
net 12	98.6	96.0	78.4	76.2	**96.0**	93.4
net 13	**99.0**	96.7	78.0	76.3	95.7	93.5
net 14	95.2	94.0	76.9	75.2	93.3	92.5
net 15	96.6	87.4	75.9	73.4	93.3	84.2

structures. The best seizure detection results were obtained for the network **net 13**. However, the best average was achieved for less complex network **net 10**. Both model consist of 4 LSTM layers but **net 10** includes a smaller number of parameters (120 contrary to 250 in case of **net 13**). Further increasing of the number of LSTM layers did not assure better results. Classification results for sequences rated as normal ones (**Criterion 2**) are entirely different. The best results are observable for the simplest networks (**net 3**). Increasing complexity of the model leads to worse results. Taking into account the overall classification quality (**Criterion 3**), the best results achieved the model **net 11**. Again a model with 4 LSTM layers proved its high performance. However, it should be stressed that there are very small differences (ca 1.5%) between models from **net 9** to **net 13**. Therefore, the model **net 9** was chosen as a trade-off between the quality of seizure detection and a number of layers and parameters.

4.2 On-line Epilepsy Detection

To definitely evaluate the quality of the developed classifier it was applied in the on-line mode to analyze the entire EEG record of every patient separately. For each patient a neurologist pointed out a particular number of seizures. For exact number of seizures the interested reader is referred to the previous authors' work [13]. The moving window of the length of 3 s was used to extract sub-sequences of EEG

Table 3 On-line sequence-to-sequence classification, epileptic patients

No.	tpr (%)	tnr (%)	acc (%)	No.	tpr (%)	tnr (%)	acc (%)	No.	tpr (%)	tnr (%)	acc (%)	No.	tpr (%)	tnr (%)	acc (%)
1	100	99.9	99.9	27	100	100	100	53	100	100	100	79	66.7	100	99.8
2	100	100	100	28	100	100	100	54	100	100	100	80	100	100	100
3	100	100	100	29	100	100	100	55	60	100	99.8	81	100	100	100
4	100	100	100	30	100	99.9	99.9	56	80	100	99.9	82	100	100	100
5	100	100	100	31	100	100	100	57	100	100	100	83	100	100	100
6	66.7	100	99.9	32	100	100	100	58	100	100	100	84	50	100	99.9
7	100	100	100	33	100	100	100	59	85.7	100	99.9	85	100	100	100
8	100	100	100	34	100	100	100	60	100	100	100	86	100	100	100
9	50	100	99.9	35	80	100	99.9	61	80	100	99.9	87	83.3	100	99.9
10	100	100	100	36	100	100	100	62	100	100	100	88	100	100	100
11	100	100	100	37	100	100	100	63	100	100	100	89	100	100	100
12	100	100	100	38	100	100	100	64	100	100	100	90	80	100	99.9
13	100	100	100	39	100	100	100	65	100	100	100	91	100	100	100
14	100	100	100	40	100	100	100	66	100	100	100	92	100	100	100
15	100	100	100	41	100	100	100	67	100	100	100	93	100	100	100
16	100	99.9	99.9	42	100	100	100	68	100	100	100	94	100	100	100
17	100	100	100	43	100	100	100	69	100	99.9	99.9	95	100	100	100
18	100	100	100	44	100	100	100	70	100	100	100	96	100	100	100
19	100	100	100	45	83.3	100	99.9	71	66.7	100	99.9	97	80	100	99.9
20	100	100	100	46	100	100	100	72	80	100	99.9	98	100	100	100
21	100	100	99.9	47	50	100	99.9	73	100	100	100	99	100	100	100
22	80	100	100	48	83.3	100	99.9	74	100	99.9	99.9	100	100	100	100
23	100	100	100	49	87.5	100	99.9	75	66.7	100	99.9	101	100	100	100
24	100	100	100	50	100	100	100	76	87.5	100	99.9	102	100	100	100
25	100	100	100	51	100	100	100	77	60	100	99.8	103	100	100	100
26	100	100	100	52	100	100	100	78	100	100	100	104	100	100	100

record. Each extracted sub-sequence was classified using **Criterion 1**. Based on the classification results of all sub-sequences, each EEG record was evaluated using performance indexes introduced in Sect. 2. The results are reported in Table 3. The most important results are that in every case the epilepsy was diagnosed. For 82 patients the system detected all seizures. For 19 patients the system detected all seizures excluding one. In case of 3 patients the system did not detect two seizures. What is even more important, the system kept excellent results for classification of sequences marked as the normal ones but acquired from epileptic patients. For 100 patients *tnr* was equal to 100% and in case of 4 patients this rate was 99,9%. Summarizing, the obtained results are pretty good.

Table 4 On-line sequence-to-sequence classification: healthy subjects

Subject no.	acc (%)	Subject no.	acc (%)	Subject no.	acc (%)	Subject no.	acc (%)
1	54.1	20	86.1	38	88.1	56	99.9
3	82.7	21	98.4	39	80.8	57	94.7
4	35.7	22	53.6	40	95.8	58	72.4
5	87.3	23	99.9	41	94	59	96
6	56	24	86.1	42	97.3	60	100
7	99.9	25	99	43	61.3	61	95.1
8	98.8	26	93.6	44	37	62	86.9
9	71.4	27	90.5	45	91.6	63	99.7
10	93.5	28	93.5	46	95.3	64	97.2
11	99.6	29	49.6	47	99.2	65	98.7
12	82.5	30	66.7	48	73.1	66	92.7
13	99.4	31	39.6	49	45.2	67	94.1
14	56.6	32	77.6	50	97.3	68	97.7
15	78.4	33	87.7	51	95.9	69	99.1
16	98.6	34	99.5	52	94.7	70	93.6
17	87	35	73.2	53	96.7	71	99.8
18	69.9	36	99.5	54	99.9	72	94.1
19	99.9	37	92	55	57.6		

4.3 System Evaluation—Healthy Subjects

The final test of the proposed system was the performance evaluation on the group of 71 healthy subjects. The results are listed in Table 4. The specificity index (in case of healthy subjects the specifity is identical to the total accuracy) takes the values from 35.7 to 100%. In case of 41 examined subjects *acc* was greater than 90%. It means that Type I classification errors are very small.

5 Sequence-to-Label Classification

5.1 Model Developing

In this scheme, the data set containing 588 epileptic samples and 588 samples acquired from healthy subjects was enriched with 1040 samples acquired from epileptic subjects but marked as the normal operation. The idea behind this is to improve the proper classification both epileptic and normal states. The formed data set was split into the training and testing sets. Both sets were separable one from another.

Table 5 Sequence-to-label: investigated structures

Network	Number of nodes			
	1st layer	2nd layer	3rd layer	4th layer
net1	100	–	–	–
net2	50	50	–	–
net3	50	100	–	–
net4	100	70	–	–
net5	10	30	50	–
net6	50	30	10	–
net7	50	50	50	–
net8	50	100	150	–
net9	100	70	40	–
net10	50	50	50	50
net11	70	60	50	50

Each examined classifier was trained 10 times using the repeated hold-out method with the dividing ratio equal to 0.5. The input space was equal to 16. This time our goal was to classify the input sequence to either "epileptic" or "healthy" class (see Fig. 2b). As we deal with short-time seizures we used input sequences of the length of 1 s. Again, we investigated different neural network structures shown in Table 5. For all structures σ_c was selected as the hyperbolic tangent function, σ_s was the sigmoid function, and σ_h was the linear one. Each neural network was trained using the ADAM algorithm with the initial learning rate set to $\eta = 0.02$. After training, each model was evaluated using tpr, tnr and acc indexes. We used also averaged indexes ($\overline{tpr}$, $\overline{tnr}$ and $\overline{acc}$, respectively) derived over 10 trials. The model evaluation is shown in Table 6 where the best results are marked by the bold face. Analysing the averaged results the model **net 5** achieved the best values of both the sensitivity as well as the overall accuracy. Concerning the averaged specificity the best results are observed for the model **net 10**, however the differences between models are not significant. In turn analysing results obtained for the best performing networks the best results for tpr and acc are seen for the model **net 7**. Again, the best tnr index was obtained for the model of the structure **net 10**. As the main objective is to detect epileptic seizures the model **net 7** was selected for further experiments.

5.2 On-line Epilepsy Detection

The developed classifier was also applied in the on-line mode to analyze the entire EEG record of every patient separately. This time the moving window was the length of 1 s. Each extracted sub-sequence was classified and then the EEG record was evaluated using performance indexes introduced in Sect. 2. The results are presented

Table 6 Evaluation results

Network	$\overline{tpr}$ [%]	$\overline{tnr}$ [%]	$\overline{acc}$ [%]	tpr_{best} [%]	tnr_{best} [%]	acc_{best} [%]
net 1	73.3	90.4	86.1	73.3	90.8	86.1
net 2	73.2	92.4	87.3	76.2	94.3	88.5
net 3	71.6	92.3	86.7	76.2	93.6	88.2
net 4	73.0	92.5	87.3	78.1	94.4	88.1
net 5	**76.0**	92.8	**88.3**	79.6	94.1	89.3
net 6	74.5	92.7	87.8	76.2	94.3	89.2
net 7	74.1	93.0	87.9	**79.8**	94.0	**89.8**
net 8	72.8	92.1	86.9	78.9	93.6	88.6
net 9	73.8	92.8	87.7	79.1	94.6	89.4
net 10	74.2	**93.3**	88.2	78.4	**95.1**	89.4
net 11	75.0	92.5	87.8	79.4	93.3	88.9

in Table 7. In every case the epilepsy was diagnosed with tpr reaching the value from 20 to 100%. For 33 patients the system detected all seizures pointed out by a neurologist. In turn, the specifity index took the values from 27.7 to 100%. It means that in some cases the system generated a large number of false alarms about seizures. Summarizing, the overall accuracy was from the interval 28.2–99.9%. Comparing these results to those presented in Table 3 we can clearly state that the sequence-to-sequence model performed much better that the sequence-to-label one.

5.3 System Evaluation—Healthy Subjects

The last examination was checking the performance of the system on the group of healthy subjects. Results are presented in Table 8. The total accuracy took the values from the interval [68%,100%]. In case of 66 examined subjects acc was greater than 90%. Comparing the achieved results to those presented in Table 4 we can clearly state that this time the sequence-to-label model performed much better that the sequence-to-sequence one.

5.4 Classifier with Dropout

We also undertook an attempt to improve the generalization of the best performing classifier **net 7** by means of dropout method. Then, the structure shown in Fig. 3 was equipped with dropout layers placed after each LSTM processing layer. The probability of dropout was set to $p = 0.2$. The results of model training are presented

Table 7 On-line sequence-to-label classification, epileptic patients

No	tpr (%)	tnr (%)	acc (%)	No.	tpr (%)	tnr (%)	acc (%)	No.	tpr (%)	tnr (%)	acc (%)	No.	tpr (%)	tnr (%)	acc (%)
1	50	62.5	62.5	27	100	88.9	88.9	53	100	56.9	57	79	83.3	76.8	76.9
2	85.7	87.8	87.8	28	100	68.3	68.6	54	50	95.9	95.6	80	83.3	77.4	77.4
3	100	84	84.1	29	60	68	68	55	100	85.1	85.2	81	100	69.6	69.9
4	100	90	90	30	100	63.6	63.9	56	80	100	99.9	82	40	93.8	93.6
5	83.3	84.8	84.8	31	80	94.9	94.8	57	100	82.1	82.2	83	100	95.8	95.8
6	66.7	98.3	98.2	32	66.7	84.2	84.1	58	100	72	72.2	84	50	97.6	97.5
7	75	81.6	81.6	33	80	37.5	37.8	59	85.7	86.5	86.5	85	63	92.6	92.30
8	100	84	83.6	34	100	68.4	68.6	60	100	77.5	77.6	86	100	56.2	56.6
9	50	98.6	98.6	35	60	95.7	95.5	61	60	84.1	84	87	67	80.7	80.7
10	100	61.2	61.4	36	83.3	86.7	86.7	62	57.1	95.2	94.9	88	60	94.1	94
11	83.3	77.4	77.4	37	80	75.7	75.7	63	83.3	90	89.9	89	50	99.6	99.5
12	75	95.6	95.4	38	80	86.9	86.9	64	77.8	89.7	89.6	90	60	90.1	89.9
13	66.7	91.2	91.1	39	50	85.7	85.6	65	60	90.2	90	91	80	90.3	90.2
14	75	71.9	71.9	40	50	83	82.8	66	100	65.6	65.7	92	20	87.4	87.1
15	85.7	61.2	61.4	41	25	96.7	96.3	67	86	67.8	68	93	75	67.7	67.7
16	90	27.7	28.3	42	80	85.7	85.7	68	100	56.1	56.2	94	75	65.1	65.1
17	75	79.1	79.1	43	80	92.6	92.5	69	100	99.5	99.5	95	100	73.3	73.5
18	100	58.6	58.9	44	40	89.5	89.2	70	88	54.9	55.2	96	88	77	77.1
19	88	78.4	78.5	45	66.7	93.2	93	71	100	95.6	95.6	97	100	96.2	96.2
20	83.3	96.8	96.7	46	100	81.8	82	72	60	89.8	89.6	98	100	82.8	82.9
21	60	99	98.8	47	100	99	99	73	88.2	91.8	91.8	99	100	59.7	60.2
22	100	60.8	61	48	66.7	67.4	67.4	74	100	74.4	74.4	100	67	59.7	59.7
23	66.7	77.3	77.2	49	75	59.2	59.3	75	100	96.2	96.2	101	75	87	87
24	57.1	81.4	81.3	50	80	72.9	72.9	76	88	95.2	95.2	102	100	27.7	28.2
25	80	89.5	89.4	51	50	81.7	81.5	77	40	98.7	98.4	103	25	94.8	94.5
26	80	71.6	71.6	52	90	57.6	57.8	78	100	76.5	76.8	104	100	97.6	97.6

in Table 9 where the best results are marked by the bold face. Clearly, using dropout we can improve the generalization of the classifier. In predominant number of cases classifier with dropout outperform the normal one.

We compared also both models on the data sets of epileptic patients and healthy subjects. The evaluation results are shown in Table 10. Again, the best results are marked by the bold face, where $\widetilde{tpr}$, $\widetilde{tnr}$ and $\widetilde{acc}$ represent the average sensitivity, specifity and total accuracy, respectively, calculated over the entire population of examined people. It is obvious that the dropout can significantly improve the classification results. We can achieve a couple of percent of improvement in the case of epileptic patients. For healthy subjects, the improvement is not so meaningful but yet it is.

Table 8 On-line sequence-to-label classification: healthy subjects

Subject no.	acc (%)	Subject no.	acc (%)	Subject no.	acc (%)	Subject no.	acc (%)
1	90.4	20	95.1	38	98.5	56	99.9
3	88.9	21	99.8	39	98.2	57	98.2
4	79.1	22	92.3	40	98.9	58	91.7
5	96	23	100	41	98.8	59	99.6
6	71.7	24	97.6	42	99.2	60	99.9
7	99.8	25	99.7	43	94.9	61	98.2
8	99.8	26	99.2	44	94.3	62	95.6
9	94.9	27	96.2	45	98.1	63	99.9
10	98.2	28	99.7	46	97.6	64	99.2
11	99.7	29	93.4	47	99	65	99.7
12	97.8	30	85.2	48	94.2	66	98.2
13	99.9	31	68	49	97.5	67	98.3
14	95.4	32	92.6	50	99.1	68	99.2
15	95.2	33	97.4	51	98.8	69	99.4
16	99.7	34	100	52	98.5	70	98.2
17	97.2	35	92.6	53	97.6	71	100
18	92.1	36	100	54	100	72	96.3
19	99.7	37	98	55	92.2		

Table 9 Model comparison—training

Network	$\overline{tpr}$ [%]	$\overline{tnr}$ [%]	$\overline{acc}$ [%]	tpr_{best} [%]	tnr_{best} [%]	acc_{best} [%]
net 7	74.1	**93.0**	87.9	79.8	94.0	89.8
net 7 with dropout	**76.7**	92.7	**88.4**	**81.3**	**94.4**	**90.2**

Table 10 On-line classification

Model	Epileptic patients			Healthy subjects
	$\widetilde{tpr}$	$\widetilde{tnr}$	$\widetilde{acc}$	$\widetilde{acc}$
net 7	**78.7**	80.5	80.5	94.7
net 7 with dropout	**78.7**	**84.1**	**84.1**	**95.4**

6 Concluding Remarks

This work reports the application of deep LSTM networks to develop computer-aided system for diagnosing short-time epileptic seizures. Two classification settings were proposed and analyzed: the sequence-to-sequence mapping and sequence-to-label classifier. Analysing the achieved results we can conclude that the proposed approaches perform pretty well. However, the sequence-to-sequence model outperform the sequence-to-label one for the seizure detection tasks. In turn, the sequence-to-label model works much better for classifying the healthy cases. The further research effort can be directed to combine both models in order to develop a classifier ensemble. Undoubtedly, the proposed system may be a valuable and very fast alternative to the visual inspection method.

References

1. Durka, P.: Adaptive time-frequency parametrization of epileptic spikes. Phys. Rev. E **69**, 051914-1–051914-5 (2004)
2. Engel, J.: Seizure and Epilepsy. FA, Davis, Philadelphia, PA, USA (1989)
3. Guo, L., Rivero, D., Dorado, J., Rabunal, J.R., Pazos, A.: Automatic epileptic seizure detection in EEGs based on line length feature and artificial neural networks. J. Neurosci. Methods **191**(1), 101–109 (2010)
4. Hochreiter, S., Schmidhuber, J.: Long short-term memory. Neural Comput. **9**, 1735–1780 (1997)
5. Hopfengartner, R., Kerling, F., Bauer, V., Stefan, H.: An efficient, robust and fast method for the offline detection of epileptic seizures in long-term scalp EEG recordings. Clin. Neurophysiol. **118**, 2332–2343 (2007)
6. Kahn, Y., Gotman, J.: Wavelet based automatic seizure detection in intracerebral electroencephalogram. Clin. Neurophysiol. **114**, 898–908 (2003)
7. Kingma, D., Ba, J.: Adam: a method for stochastic optimization. IEEE. arXiv:1412.6980 (2014)
8. Koziarski, M., Cyganek, B.: Impact of low resolution on image recognition with deep neural networks: an experimental study. Int. J. Appl. Math. Comput. Sci. **28**, 735–744 (2018)
9. Krizhevsky, A., Sutskever, I., Hinton, G.E.: Imagenet classification with deep convolutional neural networks. Paper presented at the Advances in neural information processing systems (2012)
10. LeCun, Y., Bengio, Y., Hinton, G.: Deep learning. Nature 436–444 (2015)
11. Ronneberger, O., Fischer, P., Brox, T.: U-net: Convolutional networks for biomedical image segmentation. In: Medical Image Computing and Computer-Assisted Intervention (MICCAI), vol. 9351, pp. 234–241. Springer (2015)
12. Rutkowski, G., Patan, K.: Extraction of attributes of the EEG signals based on the Stockwell transform. Pomiary Automatyka Controla **59**(3), 208–211 (2013)
13. Rutkowski, G., Patan, K.: Detection of epileptic seizures via deep Long Short-Term Memory. In: Springer Nature Switzerland (Advances in Intelligent Systems and Computing) vol. 1033, pp. 166–178 (2019)
14. Rutkowski, G., Patan, K., Leśniak, P.: Comparison of time-frequency feature extraction methods for EEG signals classification. Lecture Notes in Artificial Intelligence vol. 7895, pp. 187–194 (2014)
15. Rutkowski, G., Patan, K., Leśniak, P.: Computer aided on-line seizure detection using Stockwell transform. Adv. Intell. Syst. Comput. **230**, 279–289 (2014)

16. Schmidhuber, J.: Deep learning in neural networks: an overview. Neural Netw. **61**, 85–117 (2015)
17. Shiau, D.S., Halford, J.J., Kelly, K.M., Kern, R.T., Inman, M., Chien, J.H., Pardalos, P.M., Yang, M.C.K., Sackellares, J.C.: Signal regularity-based automated seizure detection system for scalp EEG monitoring. Cybernet. Syst. Anal. **46**, 922–935 (2010)
18. Srivastava, N., Hinton, G., Krizhevsky, A., Sutskever, I., Salakhutdinov, R.: Dropout: a simple way to prevent neural networks from overfitting. Lecture Notes in Computer Science pp. 189–204 (2014)
19. Subasi, A.: EEG signal classification using wavelet feature extraction and a mixture of expert model. Expert Syst. Appl. **32**, 1084–1093 (2007)
20. Tzallas, A.T., Tsipouras, M.G., Tsalikakis, D.G., Karvounis, E.C., Astrakas, L., Konitsiotis, S., Tzaphlidou, M.: Automated epileptic seizure detection methods: a review study. In: Stevanovic, D. (ed.) Epilepsy—Histological, Electroencephalographic and Psychological Aspects. InTech (2012). ISBN: 978-953-51-0082-9
21. Wielgosz, M., Skoczeń, A.: Using neural networks with data quantization for time series analysis in LHC superconducting magnets. Int. J. Appl. Math. Comput. Sci. **29**, 503–515 (2019)
22. Yuan Q, Zhou W, Zhang L, Zhang F, Xu F, Leng Y, Wei D, Chen M.: Epileptic seizure detection based on imbalanced classification and wavelet packet transform. Seizure **50**, 99–108 (2017)
23. Z-Flores, E., Trujillo, L., Sotelo, A., Legrand, P., Coria, L.N.: Regularity and Matching Pursuit feature extraction for the detection of epileptic seizures. J. Neurosci. Methods **266**, 107–125 (2016)

Patient Managed Patient Health Record Based on Blockchain Technology

Konrad Zaworski and Marcin Szpyrka

Abstract Patients would like to be at the center of the entire information exchange process between medical entities and gain control over their data and further medical decisions. The use of blockchain technology may create this possibility. The work describes the most important aspects of the blockchain technology and digital medical records. The challenges faced by architects of PHR systems based on blockchain are presented. Examples of theoretical and existing solutions are examined and their weaknesses are quoted. We propose a new solutions to this issue. Our main goal is to preserve the characteristics of blockchain without robbing it of its most important advantages.

Keywords Patient Health Record · PHR · Blockchain · Bitcoin

1 Introduction

In October 2008, an unknown person or group of people nicknamed Satoshi Nakamoto published a document on the Internet entitled "Bitcoin: A Peer-to-Peer Electronic Cash System" [1], being the first conceptualization of blockchain technology, a decentralized and distributed database. The publication described the possibility of implementing an electronic currency (cryptocurrency) that would not have a central entity supervising the circulation of coins and their production. The solution was supposed to guarantee freedom from any state influence, eliminate the problem of printing money and a sudden change in the rules of use and ownership. Bitcoin removes these restrictions because it is based on cryptographic evidence of transactions that assures their occurrence and irreversibility. From a small socio-economic

K. Zaworski (✉) · M. Szpyrka
Department of Applied Computer Science, Faculty of Electrical Engineering, Automatics,
Computer Science and Biomedical Engineering, AGH University of Science and Technology,
al. Mickiewicza 30, 30-059 Kraków, Poland
e-mail: zaworski@agh.edu.pl

© The Author(s), under exclusive license to Springer Nature Switzerland AG 2021   173
J. Korbicz et al. (eds.), *Advances in Diagnostics of Processes and Systems*,
Studies in Systems, Decision and Control 313,
https://doi.org/10.1007/978-3-030-58964-6_13

experiment, Bitcoin has now become a global means of payment, looking to obtain legal status of the currency.

Bitcoin started a revolution not only in economic area. The Blockchain technology on which it is built, finds its origins in the 90s, when S. Haber and W. S. Stornetta described a cryptographic system, time-stamped document storage [2], whose integrity of the set could not be compromised. One year later they extended their concept [3] by supplementing it with the use of Merkel tree. Thanks to this, the verification of a single document could take place without having the entire collection.

Special attention of the scientific community is drawn to the use of blockchain technology in medicine. The 21st century is a time of dynamic development of medical systems, which goal is to exchange information effectively between entities. Current solutions put the patient out of the entire process, not giving him the ability to manage his data. He cannot fully control who has access and where it is stored. Blockchain has the potential to revolutionize this pattern by putting the patient at its center. Then patient decides who and under what conditions will receive access to his medical information. Using techniques such as decentralization, data encryption, time stamping or distributed consensus algorithms, blockchain offers a comprehensive solution to problems related to the profitability of operations, low performance and potential security problems in centralized, medical systems [4]. Blockchain technology can be used in all places where information added to the collection must be indelible, its appearance described at a certain point in time, the integrity of the content can be verified without having the entire collection and the accuracy of contained information must be determined. This work focuses on this issue, indicating the prospects for using blockchain technology in the field of Interconnected Patient Health Record. Some of existing solutions and ideas were examined. Own proposals and new development directions in this area are presented.

The paper is organised as follows. Section 2 contains an introduction to blockchain technology. It presents the basic elements included in the blockchain network and its techniques providing decentralization, time stamping and distributed consensus algorithms. Section 3 describes Patient Health Records systems, their advantages, requirements and architectonical structures. Sections 4 and 5 give general outlook how blockchain can be introduced into PHR systems, examples of such projects and proposals about improving cooperation between these two technologies. A short summary is given in the final section.

2 Blockchain Technology

Blockchain consists of blocks, organized in the order they are added, linked together using hashes calculated from their content and the hash of the previous block. Each block contains a differentiated number of transactions that store information relevant to the database. In the case of Bitcoin, this data is describing flow of tokens (coins) on the network. Transactions are added to the collection by entities (individuals and companies) represented in the database in the form of a pair of keys: public, visible (directly or indirectly) to other entities, and private—used by people to confirm performed operations in their own names. Transactions are grouped into a block and added to the chain after verification, which is carried out by entities called miners. They are people who use their server resources to cryptographically sign data—for which they receive payment from the network, e.g. in the form of bitcoins (coins) generated to their address.

Each user performing blockchain operations has a self-generated private key. Thanks to it, it is possible to prove being the author of transactions. Based on the private key, it is possible to obtain a complementary public key. Because addresses are long enough, random strings, it is extremely unlikely that two people would generate the same address. These situations are called collisions and no case has been documented yet that would lead to the theft of data assigned to the address.

Transactions are a data carrier in the blockchain. They don't have to, but they can be related to each other, which can be useful, for example, in the transfer of tokens. In Bitcoin, they are responsible for transferring coins from one address to another. Each transaction has any number of inputs and outputs. Inputs are references to the outputs of previous transactions, from which received funds are forwarded. Outputs contain instructions for sending funds to other addresses. The sum of funds at outputs must be equal to the sum at inputs. Tokens stored in specific address (public key) can be easily calculated by tracking transactions outputs.

The original idea of blockchain assumed its usefulness in the circulation of documents [2, 3]. The proposed solutions assumed that the documents (in the case of Bitcoin these are transactions) are grouped into subsets called blocks.

Blocks are linked together to form a chain that cannot be broken or modified. This is because the hash of each block is calculated from its metadata and the hash of the previous block (Fig. 1). By modifying data in a block, its hash automatically changes, preventing chain verification from being successfully completed. To optimize the document verification process, Merkel trees have been used. The algorithm consists of calculating the hashes of the documents or transactions included in the block, and then the obtained hashes are combined in pairs at a higher level and recalculated, until one hash representing the entire subset of documents is obtained. A blockchain client who does not have a full chain verifies the existence of a document or transaction by downloading only block headers and then asks different

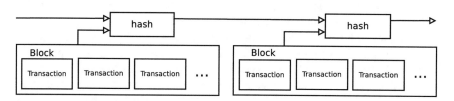

Fig. 1 Blocks connected by hashes [1]

network members about Merkel proof (hashes required for verification). Replacing any transaction in a block would change all hashes at higher levels, and the hash of the entire block.

Adding a new block to the chain in the blockchain network consists of its verification. People called miners use their resources, such as their computing power, to achieve benefits, e.g. in the form of network tokens, which they can later sell for fiat money. In order to avoid a situation where many network members want to add blocks of different content at once, or add a block that does not comply with accepted standards, blockchain uses one of the consensus algorithms [5]. Obtaining a consensus means that all or most of the network members agree to add a given block and it becomes an integral part of the blockchain from that moment.

3   Interconnected Patient Health Records

Supported by Internet trends and ideas promoting patient rights, the social need for flexible access to medical information and services is growing. People get health information from many sources, including television, the Internet, doctors and friends. There is a natural need to make this information a natural part of the healthcare system. The future is closely related to sharing medical data with patients who, according to research [6], would like to be at the center of the entire information exchange process between medical entities and gain control over their data and further medical decisions.

One of the types of systems implemented in medical care facilities is Patient Health Record. In the basic scope, PHR allows the patient to access his own health card. The goal is to interact with medical data through a secured communication channel. It can allow patients to complete, view and share their medical information. This does not necessarily mean that the patient receives access to all his data and controls how it is transferred to other entities. The term PHR in the scientific literature is flexible and is often extended with additional improvements [7]: providing medical care via the Internet, checking test results, verification of the information entered by patients, tracking of costs information, printing documents, secure messaging, online prescriptions, booking visits via the Internet, access to educational materials, reminders and medical alerts or even address books.

From the architectural point of view, PHR systems can be divided into three types: Stand-alone PHR, Tethered PHR and Interconnected PHR [8].

Stand-alone PHR is the simplest architectural form of PHR. The data is entered by the patient and saved on a USB stick, CD, smart card, in a desktop application or available in the form of a website. In this model, the patient has full control over the data, but there is no connection with other systems of this type. Therefore, the quality of the data is questionable. This approach can be compared to paper records kept by patients at home.

In the Tethered PHR model, systems are maintained and made available to patients by medical facilities. The documentation is completed by the qualified staff, and

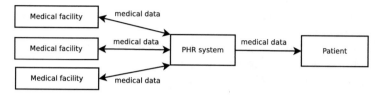

Fig. 2 Model interconnected PHR

a patient has little influence on the data stored in the database. What patient sees can be limited in accordance with the policy of the institution—they can be, for example, only prescriptions and test results [9]. Each medical facility has its own PHR system. The patient has access to this data—usually only in read mode.

The solutions discussed in this paper can be qualified as elements of the Interconnected PHR system, because the patient can acquire full control over his data and supervise exchange of his information between medical institutions. The implementation of such a model is complicated due to the possibility of transporting medical information from one facility to another and data from many sources is made available to the patient. In this solution, it is important to answer the question where the information is located and who bears the costs of its storage. Patients are not specialists in the field of databases, they do not have technological facilities in their homes, and they cannot even afford to maintain such infrastructure. Choosing a central data storage point, for example one of medical facilities, gives the patient only apparent control in the process of information exchange. Blockchain can help to solve this problem, because it maintains data in a distributed form, giving patients decision-making capabilities through cryptographic mechanisms.

Figure 2 shows a simplified Interconnected PHR model. All facilities use one common PHR system which they exchange medical information with. The patient has access to this data—usually only in read mode. It should be noted that PHR does not have to replace internal facilities systems. Many centers have their own EHR / EMR systems, while PHR serves only as a tool for integration with other private systems, ensuring the collection of patient data from all possible places in one database [10, 11].

The implementation of the Interconnected PHR system, which is often a long and expensive process, requires the presentation of solid arguments in favor of this solution and eliminating the disadvantages arising from its requirements. Based on [7, 8, 12–14] there are many benefits coming from PHR systems:

- Patients can be identified in a unique way in all medical facilities.
- Unification of the terminology ensures universality and readability of data.
- All information generated in one of the medical points can be immediately made available to other facilities.
- Access to medical data is not limited to where it was generated.
- It is possible to process the entire data set for better therapy selection for current and future patients and for statistical purposes.

- Providing staff with high-quality, precise data translates into increased quality of healthcare.
- Problems related to storing large amounts of data and their low durability in the case of traditional writing on paper are eliminated.
- Patients can verify data in the system and confirm that it is complete.
- Online booking save the time of patients and staff.
- Access to own documentation and educational materials allows patients to understand their health and make better decisions about therapies.
- Parts of documentation can be copied into other documents multiple times.
- Medical staff not only gain integrated access to all patient data, but also receive different views of this information (e.g. chronological or filtered by type of event).
- Due to the precision of the documentation and automatic analysis of the collected data by the system algorithms, it is possible to avoid medical errors that are dangerous to the patient's health.
- Performance of facility is increased while operating costs are reduced.

4 PHR Based on Blockchain

The designers of block-based PHR systems also face a number of challenges that may be the key to the profitability of implementing such a solution. The processing speed of the blockchain can be slow due to signature verification, obtaining consensus and data redundancy [15]. It is worth considering how to exclude certain blockchain mechanisms or minimize them so that the data appears in the chain as soon as possible. Block verification in the Bitcoin network requires significant computing power due to the consensus algorithms used. The electricity consumed in this process is wasted and from the ecological point of view the verification process in PHR systems should be improved. Blockchain data flows through all nodes. It is also easy to track which transactions are associated with a given public key. A key pair can be used to encrypt part of the data, but additional tracking protection mechanisms should be implemented.

For PHR systems, data size is a key factor affecting the entire project. Individual imaging results can take up to several gigabytes. A system supporting millions of patients cannot use full data redundancy because the storage costs would be too big, and the possibility of replication exabytes of data to all facilities is negligible. Instead, the off-chain storage model can be used, which assumes storing part of the data in the datacenters of the facilities (not in blockchain) and regulated access to them with the consent of the patient.

The scientific literature describes the cases of using blockchain technology in PHR systems, however, the vast majority break the assumptions of the blockchain model to eliminate design flaws. A good question is whether they are still solutions using blockchain.

In [15] secret-data sharing solution was proposed. There are five entities: the owner, the user, the gateway server, the cloud storage and the private blockchain. The owner owns secret-data, reads and writes it to the cloud server through the gateway. He controls access to his information by having encryption key and storing secret-data metadata and permissions in blockchain. The user is someone who wants access to secret-data located in the cloud storage. It finds interesting data in blockchain and requests access from the owner. The gateway server verifies entities and checks permissions using blockchain and re-encrypts secret-data from cloud storage. This model assumes that only predefined list of users can access private blockchain (for example medical entities), because it contains non-encrypted metadata about data stored in the cloud. There is also central server called gateway, which could break the idea of blockchain technology. A good question is also whether in this solution blockchain could be replaced by a standard, relational database if the records are only added and modified by the gateway.

OmniPHR [16] divides patient information into small pieces which are represented as blocks in the blockchain. This solution uses Chord algorithm in order to ensure equal distribution and knowledge of nodes where datablocks are located. Each datablock is encrypted and cryptographically signed by the responsible for inserting the information—a health professional, patient or whom the patient authorized access their health records. Due to architectural reasons model needs to store data on the node closest to the user, for example data created by a physician in a hospital is stored in the datacenter of facility with some copies of these datablocks distributed over the network. What is more, as default data is shared only between healthcare provider and patient (sharing patient data between facilities may be difficult).

MediBloc [17] project stores metadata in the blockchain and real patient data in the Distributed Hash Table (DHT) using Inter Planetary File System (IPFS). It is based on Qtum public blockchain, which assumes existence of transaction fees and usage of POS (Proof of Stake) consensus algorithm, which could be unefficient. Model assumes, that medical data will be stored in patient devices like personal computers or smartphones (with optional backup in MediBlock project storage). What is more, data is encrypted and decrypted only using patient key pair which leads to question if data could be effectively shared with facilities. All these disadvantages mean that this solution could have the performance, scalability and energy consumption issues [15].

MedVault [18] uses new types of cryptographic mechanisms called redactable signature scheme, developed by project team, as a core component. It permits patients to selectively disclose information from their PHRs to third parties while preserving integrity and source verifiability of the data. In this approach medical facilities supply a third party with records such as doctors' notes, lab results and diagnoses. Requests for records are sent to the third party, which follows an authorization policy set for each patient before data is released from its database. Medical records are directly stored in the blockchain which is assumed to be privacy-preserving, but project team have not yet published their blockchain-based design [19].

MedRec and MedShare [15, 20, 21] are similar and both use feature called smart contracts to exchange data between medical facilities under control of patient. They use Ethereum public blockchain which is strongly connected with transaction fees and involves users into block verification work (mining) in order to obtain tokens required for participation in the network. Performance, scalability and energy consumption are the main issues that may put into question the usefulness of these projects in PHR solutions.

In [22] authors propose to build a new layer on top of the blockchain in order to connect data for one specific patient. Obtaining all records for only one patient could be problematic due to orderly nature of blockchain, where blocks containing data about milions of users would be connected in a straight line. Creating additional chains for every user can be efficient but may reveal too much information about one person. For this reason, authors introduced a different way of connecting blocks in the upper layer, in which the hash of the next block is calculated from the hash of the previous block and the salt known only to the patient. This way only owner of these blocks knows how to traverse them. Authors also propose using of Storj and InterPlanetary File System (IPFS) projects in order to split data in the network into small chunks. This solution, however, requires clarifying who will bear the costs of redundancy and how to protect against intentional, excessive and malicious redundancy.

5 Proposed Solutions

The solutions presented in the previous chapter have disadvantages. Some of them are significant enough to ask if these projects still benefit from the idea of blockchain. Our goal is to achieve the perfect combination of blockchain technology with a PHR system while eliminating the disadvantages of protocol restrictions. The solutions presented in this chapter are only a general outline of loosely related concepts that we plan to explore and improve in the future.

We focus on a liberal solution in which the patient decides who and under what conditions has access to his medical data. At the same time, data may be added by third parties on behalf of the patient with his consent. He cannot store data only on his private devices and his medical information can be accessible from any place in the world. This has to be a national solution in which one of the entities is the ministry of health.

The first step in developing a Blockchain-based PHR model is to determine what type of blockchain access we are dealing with. The three main types are private, public and consortium [15]. Due to the possibility of saving data in blockchain by both medical facilities and patients, the most accurate choice seems to be the consortium model. This is a semi-private blockchain managed by many entities. This solution is suitable for cooperation of various companies while maintaining security and rules prevailing in public blockchain. The list of network participants is explicitly declared and only they have access to it. A predefined group of entities could grant access to the network to new facilities and patients. For example, a national insurer or a health ministry could decide who on the network is an institution and who is a patient. In this way we limit the number of participants in the model and protect the network from

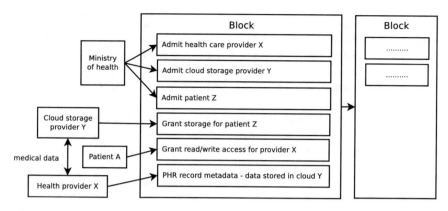

Fig. 3 Simplified solution model

flooding with insignificant data. In proposed solution there are 4 entities: the health ministry, health care providers (including IoT devices), patients and cloud storage providers (Fig. 3).

The ministry of health announces in the form of transactions who is a health provider, patient and cloud storage by broadcasting their public keys. In the same way it can revoke access. It signs relevant agreements with cloud providers who will store medical records on their servers. Providers take care of data redundancy according to strictly defined rules. At the request of the ministry, the cloud provider announces on the network in the form of transactions that it supports specific patients. There may be more super members in the model who can admit entities.

The above solution may raise questions about breaking the rules of decentralization in Blockchain. Roles are assigned by the Ministry of Health, which means that there is an entity with special rights over other network users. It is worth noting, that this is not a public solution, but a semi-private, managed by many entities. Furthermore, the subject of blockchain decentralization are not permissions but data. In the proposed solution, the data is not kept in one place and none of the entities acts as an access gateway to information. Data can be replicated over many cloud providers ensuring constant and equal access for all entities. In this case, the rule of permissionless has been broken, however, this is fully acceptable in semi-private blockchains.

The patient authorizes the facility to access his data stored in the cloud. A medical facility that wants to add a new patient record is looking in the blockchain which cloud providers support the patient and sends data to one of them. It also adds the record's metadata and location information to the chain.

There are many types of transactions in the proposed model, some are responsible for granting access, others only confirm the existence of medical records in the cloud. Therefore, the additional blockchain layers mentioned in the previous section are needed so that data about one patient is downloaded efficiently and saved in a way that maximally anonymizes his information [22].

Unlike the solutions presented in the previous section, medical facilities are not forced to store data of various patients, including those who do not use their services. Instead, the ministry ensures storage through appropriate agreements with cloud providers. In addition, the model can be extended with additional contracts signed by individual medical facilities to provide them with preferential access to data, while not cutting off access to other facilities.

An important element of the system is encryption of record metadata in blockchain. We are considering a solution in which there are temporary chains in the network, where, for the needs of a given facility, the patient using application decodes his record transactions and encrypts them using the institution's public key. The data is kept there for a specified period of time until it is collected by the facility.

The last element to discuss is block verification. To avoid centralization, this process cannot be carried out by the ministry itself. Assuming that we are dealing with consortium blockchain, verification can be done by medical facilities using a modified PoS algorithm, which is based on the number of patients served in a given time window. The ministry could offer various types of rewards for institutions for verified blocks.

The proposed solution is general, it does not answer directly to more detailed questions, for example about the resignation of one of the cloud providers and the transfer of data to another one. We intend to develop the given solution and specify unknown issues in subsequent publications on this subject.

The Table 1 shows comparison of the projects described in the previous chapter with the proposed solution.

Table 1 Comparison of solutions presented in the described publications

	Proposed solution	Pub. [15]	OmniPHR [16]	MediBloc [17]	MedVault [18]	MedRec and MedShare [20, 21]
Justified blockchain implementation	✓	✗	✓	✗	✓	✓
Decentralization of data	✓	✗	✓	✗	✓	✓
Data stored off-chain	✓	✓	✓	✓	✗	✓
Access to data controlled by patient	✓	✓	✓ with limitations	✓	✓	✓
Low energy consumption consensus alg.	✓	n/a	n/a	✓	n/a	✗
Patients do not bear data storage costs directly	✓	✓	✓	✗	n/a	✗

6 Summary

International research indicates a strong need to develop interoperability standards for medical systems so that they can exchange medical information with each other. Patients also require access to this data in a way that enables them to actively participate in the treatment process. In addition, an important factor is the security of stored information and communication between entities involved in the broadly understood process of medical care [23]. Blockchain can help meet these conditions, because its assumptions perfectly match the listed features. It provides extensive mechanisms of authorization, redundancy of stored data and resistance to cyber attacks, e.g. counterfeiting, system overloading and unauthorized access. Asymmetric cryptography not only provides verification of PHR participants, but can also be used to encrypt data. It provides mechanisms for anonymizing users and information about their treatment. Interconnected PHR based on blockchain can break the access barrier, scattering data between entities, allowing them free (from anywhere), but controlled access [24]. Blockchain seems to be the golden mean in developing a protocol for collecting and exchanging medical information in a way that separates the data layer from the application layer. It requires all entities to use the same standard, and its change can only take place with the consent of most of members (and later it is automatically forced on all network participants). Stored data and the way of interacting with them allows the production of many versions of client software with various functionalities, capable of interacting with the same platform. The use of blockchain shortens the development time of such systems and decreases project costs.

References

1. Nakamoto, S.: Bitcoin: a peer-to-peer electronic cash system (2008). https://bitcoin.org/bitcoin.pdf. Accessed 01 Jan. 2020
2. Haber, S., Stornetta, W.S.: How to time-stamp a digital document. J. Cryptol. **3**, 99–111 (1991)
3. Bayer, D., Haber, S., Stornetta, W.S.: Improving the efficiency and reliability of digital time-stamping. Sequences **II**, 329–334 (1991)
4. Yong, Y., Fei-Yue, W.: Blockchain and cryptocurrencies: model, techniques, and applications. IEEE Trans. Syst. Man Cybernet. Syst. **48**, 1421–1428 (2018)
5. Antonopoulos, A.M.: Mastering Bitcoin: Unlocking Digital Cryptocurrencies. O'Reilly Media, Inc., Newton (2014)
6. Coulter, A.: Engaging Patients in Their Healthcare. Picker Institute Europe, Oxford (2006)
7. Abdulkareem Alyami, M., Song, YT.: Removing barriers in using personal health record systems. In: 2016 IEEE/ACIS 15th International Conference on Computer and Information Science (ICIS), pp. 1–8 (2016)
8. Razavi, A.: Design and development of a personal health record system for prostate cancer patient (2013)
9. Hye-Jeong, J., Namhyun, K., Hasuk, B.: Development of a personal health record system based on usb flash drive and web service. J. Korean Soc. Med. Inform. **15**, 341–350 (2009)
10. Lähteenmäki, J., Muuraiskangas, S., Leväsluoto, J.: Online electronic services for preventive and customer-centric healthcare: Experiences from PHR deployment in the Tampere region, Finland. Number 232 in VTT Technology. VTT Technical Research Centre of Finland (2015)

11. Schleeper, D.: Towards an interconnected personal health system for the Dutch healthcare sector. Barriers and facilitators to implementation. Master thesis, pp. 175–185 (2017)
12. World Health Organization, Regional Office for the Western Pacific (2006) Electronic Health Records: Manual for Developing Countries. Manila: WHO Regional Office for the Western Pacific. http://www.wpro.who.int/publications/docs/EHRmanual.pdf. Accessed 01 Jan. 2020
13. Hamilton, B., Hamilton, L.: Electronic Health Records. McGraw-Hill Higher Education, New York (2011)
14. Tang, P., McDonald, C.: Biomedical Informatics: Computer Applications in Health Care and Biomedicine. Springer, New York, NY (2006)
15. Thwin, T., Vasupongayya, S.: Blockchain based secret-data sharing model for personal health record system. In: 2018 5th International Conference on Advanced Informatics: Concept Theory and Applications (ICAICTA), pp. 196–201 (2018)
16. Roehrs, A., André da Costa, C., Righi, R.: OmniPHR: a distributed architecture model to integrate personal health records. J. Biomed. Inf. **71**, 70–81 (2017)
17. MediBloc Team.: MediBloc whitepaper 2017. https://whitepaper.io/document/176/medibloc-whitepaper. Accessed 01 Jan. 2020
18. Blough, D., Ahamad, M., Saintfort, F., Liu, L.: MedVault: ensuring security and privacy for electronic medical records. Georgia Institute of Technology, Atlanta (2011)
19. Kokoris-Kogias, E., Alp, E.C., Siby, S.D., Gailly, N., Jovanovic, P., Gasser, L., Ford, B.: Hidden in plain sight: storing and managing secrets on a public ledger. IACR Cryptol. ePrint Arch., 209 (2018)
20. Azaria, A., Ekblaw, A., Vieira, T., Lippman, A.: Medrec: using blockchain for medical data access and permission management. In: 2016 2nd International Conference on Open and Big Data (OBD), pp. 25–30, (2016)
21. Xia, Q., Sifah, E.B., Asamoah, K.O., Gao, J., Du, X., Guizani, M.: Medshare: trust-less medical data sharing among cloud service providers via blockchain. IEEE Access **5**, 14757–14767 (2017)
22. Hanley, M., Tewari, H.: Managing lifetime healthcare data on the Blockchain. 2018 IEEE SW/SCALCOM/UIC/ATC/CBDCom/IOP/SCI, pp. 246–251 (2018)
23. Kalra, D.: Electronic health record standards. Yearbook Med. Inf. **45**, 136–144 (2006)
24. Tang, P., Ash, J., Bates, D., Overhage, J.M., Sands, D.: Personal health records: definitions, benefits, and strategies for overcoming barriers to adoptions. J. Am. Med. Inf. Assoc.: JAMIA **13**, 121–126 (2006)

Printed in the United States
by Baker & Taylor Publisher Services